HALT, HASS, AND HASA EXPLAINED

Also Available from ASQ Quality Press

Failure Mode and Effect Analysis: FMEA from Theory to Execution
D. H. Stamatis

Weibull Analysis
Bryan Dodson

Internal Quality Auditing
Denis Pronovost

Reliability Statistics
Robert A. Dovich

Reliability Methods for Engineers
K. S. Krishnamoorthi

After the Quality Audit: Closing the Loop on the Audit Process, Second Edition
J.P. Russell and Terry Regel

Root Cause Analysis: Simplified Tools and Techniques
Bjørn Andersen and Tom Fagerhaug

To request a complimentary catalog of ASQ Quality Press publications, call 800-248-1946, or visit our online bookstore at http://qualitypress.asq.org .

HALT, HASS, AND HASA EXPLAINED

Accelerated Reliability Techniques

Harry W. McLean

ASQ Quality Press
Milwaukee, Wisconsin

HALT, HASS, and HASA Explained
Harry W. McLean

Library of Congress Cataloging-in-Publication Data

McLean, Harry W., 1946–
   HALT, HASS, and HASA explained : accelerated reliability
   techniques / Harry W. McLean.
      p. cm.
   Includes bibliographical references and index.
   ISBN 0-87389-489-8
      1. Reliability (Engineering) 2. Accelerated life testing. 3. Vibration
   tests. 4. Strains and stresses. I. Title.

TS173.M42 2000
620'.00452—dc21
                                                 00-038994

© 2000 by ASQ

All rights reserved. No part of this book may be reproduced in any form or by any means, electronic, mechanical, photocopying, recording, or otherwise, without the prior written permission of the publisher.

10 9 8 7 6 5 4 3 2 1

ISBN 0-87389-489-8

Acquisitions Editor: Ken Zielske
Project Editor: Annemieke Koudstaal
Production Administrator: Shawn Dohogne
Special Marketing Representative: Matthew Meinholz

ASQ Mission: The American Society for Quality advances individual and organizational performance excellence worldwide by providing opportunities for learning, quality improvement, and knowledge exchange.

Attention: Bookstores, Wholesalers, Schools and Corporations: ASQ Quality Press books, videotapes, audiotapes, and software are available at quantity discounts with bulk purchases for business, educational, or instructional use. For information, please contact ASQ Quality Press at 800-248-1946, or write to ASQ Quality Press, P.O. Box 3005, Milwaukee, WI 53201-3005.

To place orders or to request a free copy of the ASQ Quality Press Publications Catalog, including ASQ membership information, call 800-248-1946. Visit our web site at www.asq.org or http://qualitypress.asq.org .

Printed in the United States of America

Printed on acid-free paper

American Society for Quality

Quality Press
611 East Wisconsin Avenue
Milwaukee, Wisconsin 53202
Call toll free 800-248-1946
www.asq.org
http://qualitypress.asq.org
http://standardsgroup.asq.org

Table of Contents

Acknowledgements..xi
Foreword..xiii
Preface...xv

One: The Importance of High Reliability Products at Market
    Introduction—How & Why to do a HALT........................1
    Introduction...2
    An Overview of HALT.......................................2
    Comparisons of Products with and without HALT.............6
    The HALT Process..8
        Cold and Hot Step Stress............................10
        Rapid Thermal Transitions..........................12
        Vibration Step Stress...............................13
        Other HALT Stresses and Special Situations..........14
        Combined Stresses in HALT...........................14
    Verification HALT...15
    A Perspective on Implementing Corrective Action as a Result
        of HALT...16
        HALT Limits and Issues.............................16
        Using the Process...................................17
        Vibration...20
        Conclusion..21
    HALT Summarized...21
    Illustrating the Value of HALT............................23
    Some Thoughts Regarding Ruggedizing a Product Prior
        to HALT...25
    The Recording of Failures and Corrective Action...........26
    Troubleshooting Products Under Stress Conditions..........28
    Conclusion..29

Two: Highly Accelerated Stress Screen—HASS....................31
    Introduction..31
    The ABCs of Building Robust Products......................32
    Production Product Stress Screen—HASS....................33
    Why Does HASS Work?.......................................34
    Vibration...35

Rate of Change of Temperature 36
What Levels of Stresses are Appropriate?. 37
Precipitation and Detection Screens 39
    Precipitation Screens 39
    Detection Screens 40
    A Comment on HASS Profiles or Screens 41
Proof of Screen 42
    Fixture Characterization 42
    The HASS Profile 44
    Defect Detection 44
    Life Determination in Proof of Screen 45
Screen Tuning 47
Cables for HASS 48
HASS Summarized 49
Some HASS Successes................................... 51
A Word of Caution..................................... 52
Conclusion.. 53

Three: Beyond the Paradigm of Environmental Stress Screening—
Using HASA.. 55
Introduction 55
Background 56
Statistical Process Overview.............................. 58
Statistics—The System................................... 59
A Control Chart for the HASA Process 62
The Monitoring System Issues 63
Problems Uncovered through HASA 64
An Observation on Using Equation 3.1 66
Conclusion.. 66

Four: Refinements on Highly Accelerated Stress Audit (HASA) 69
Introduction 69
Background and Assumptions 70
Application of the Statistics 71
A Graphical Tool for Detecting Defect Level Changes 74
Conclusion.. 78
Introduction to an Improved HASA Process.................. 78
HASA Process Flow.................................... 79
Typical Lot Acceptance Sampling Plan 80
HASA Acceptance Sampling Plan 81

Five: The Equipment Required to Perform Efficient Accelerated
Reliability Testing 87
Overview 87
Temperature 88
    Turbulence 89
    Heating 90

Cooling 90
A Comparison of LN_2 Systems and Compressor Systems
  for HALT and HASS 90
Vibration 92
Control Systems 95
The Chamber 95
  Product Accessibility 95
  Ducting Air 96
  Work Area Audible Noise Levels 96
  Serviceability 96
  Service 96
  Maximum System Capabilities 96
  Post Sales Support 97
Auxiliary Equipment, Operator Safety, and ESD 97
Failure Analysis 98
Conclusion 99

Six: How to Sell New Concepts to Management 101
  Introduction 101
  Overview 102
    The Situation Today 102
    The Proposed Program 104
  Addressing Potential Management Concerns 106
  The Savings 108
  Conclusion 112

Seven: Some Commonly Asked Questions and Observations?. ... 113
  How Would Someone Compare ESS and HASS? 113
  What Is HALT in a Few Words? 114
  How Would One Compare Product Qualification Methods
    and HALT? 115
  Is HALT for Quality Improvement or Is It Intended
    to Replace RGT and MTBF Tests? 116
  Is There Any Merit to Subjecting a Product Far Beyond Its
    Design Specifications? 116
  What are Product Specific and Generic Stresses? 117
  If HALT and HASS Are So Great, Why Isn't Everyone
    Using HALT and HASS? 118
  Is One HALT Enough? 119
  At What Product Level Should HALT Be Performed? 119
  Who Should Be Involved with the Accelerated
    Reliability Program? 120
  Physically, Where Should HALT Be Performed? 121
  How Many Units are Required and What Can Be Done
    with Them Once We're Finished with HALT? 121
  Why a Cultural Change May Be Required in Order to Perform a
    Successful HALT 122

Can a Conventional Chamber and Vibration Table Be Used
  to Perform a HALT? 123
Are All Six Degrees of Freedom Shakers the Same?. 125
Are There Any Known Problems in Applying Product
  Temperature Ramp Rates of ≥60C° per Minute? 125
Are There Any Advantages to Performing Sequential Rather
  than Combined Stress Regimens? 125
Do HALT and HASS Just Uncover Electronic Defects? 126
Can HASS Eliminate My Production Steady State, Elevated
  Temperature Burn-In? 126
At What Levels of Temperature and Vibration Can I Consider
  the Product Robust?. 126
How Can You Justify Doing HALT on Products with a Very
  Short Field Life? 126

Appendix A: The Derivation of Equation 3.1. 129

Glossary of Terms and Acronyms 131

References 141

Additional References. 143

Trademarks and Service Marks. 144

About the Author 145

Index. 147

List of Figures

Figure 1.1   Traditional product development 4
Figure 1.2   Product development with HALT 5
Figure 1.3   An overlay of Figures 1.1 and 1.2 6
Figure 1.4   Product life without HALT 7
Figure 1.5   Product life with HALT 7
Figure 1.6   HALT margin discovery diagram 9
Figure 1.7   HALT step stress approach model 11
Figure 1.8   An example of temperature step stressing 12
Figure 1.9   HALT defects by environment 15
Figure 1.10 Thermal statistics diagram 18
Figure 1.11 Vibration statistics diagram 20
Figure 1.12 Device Defect Tracking status description .27
Figure 1.13 Graphical representation of DDT 27
Figure 2.1   SN diagram for T6 wrought aluminum 34
Figure 2.2   Surface mount transistor lifting and tearing up substrate 37
Figure 2.3   An ideal thermal profile 41
Figure 2.4   A traditional thermal profile 42
Figure 2.5   Failures by position number for 7,399 PC assemblies 46
Figure 2.6   Failures by position number for 14K PC assemblies 47
Figure 2.7   Data gathered before tuning the screen 48
Figure 2.8   The screen after performing screen tuning .49
Figure 3.1   HASA control charts for products tested 63
Figure 3.2   Control chart for all stress system and test system problems 64
Figure 4.1   Probability of failures for acceptable and rejectable production
             for N=198, α=0.05, β=0.10, p=0.01, and D=0.03 74
Figure 4.2   Probability of failures for acceptable and rejectable production
             for N=106, α=0.10, β=0.20, p=0.01 and D=0.03 74
Figure 4.3   Adjusted cumulative defects versus units tested showing
             solder process defect correction 76
Figure 4.4   Adjusted cumulative defects versus units tested 77
Figure 4.5   Adjusted cumulative defects versus units tested 78
Figure 4.6   Operational characteristic curve 81
Figure 4.7   Average run length curve 82
Figure 4.8   Operating characteristic curve 83
Figure 4.9   Average run length curve 83
Figure 4.10 HASA sampling plan for high volume products 84
Figure 4.11 HASA decision making 85
Figure 5.1   Accelerated stress system block diagram 88
Figure 5.2   OVS-2.5e HP *(Courtesy of QualMark Corp.)* 89

Figure 5.3 QRS-600V *(Courtesy of Screening Systems Inc.)* 89
Figure 5.4 Using each type system as they were designed (optimized) 92
Figure 5.5 PSD from a six-degree of freedom vibration viewed from
z-axis 94
Figure 7.1 HALT users survey 124
Figure A.1 Decision criteria 129

List of Tables

Table 1.1 Product's thermal specifications 16
Table 1.2 Product limits with proposed statistical limits 17
Table 1.3 Sigma and deviation impact on HALT thermal limit 19
Table 3.1 Examples of α and β risks with their respective Z values 60
Table 3.2 Sample size determinations for given conditions 61
Table 4.1 Sample sizes for given α, β, p, D, and CV 72
Table 4.2 Historical HASA data used to generate Figures 4.3 through 4.5 75
Table 5.1 Comparisons of compressor system and LN_2 type system 91
Table 6.1 Personnel and space requirements under existing reliability
program 103
Table 6.2 Reliability costs under existing reliability program 104
Table 6.3 The operational costs of HALT and HASS for one year with two
stress systems 106
Table 6.4 Comparisons of personnel and space requirements for present
and proposed reliability programs 110
Table 6.5 Comparisons of on-going costs for present and proposed
reliability programs 110
Table 6.6 Costs for proposed reliability program 111
Table 6.7 Summary of savings between programs 111
Table 7.1 A Comparison of ESS, HALT, and HASS 114

Acknowledgments

I would like to express my thanks to Tim Kramer, Ph.D., a statistician at HP, Vancouver Division. His efforts are greatly appreciated. I wish to thank Valerie Wildman, Ph.D. and Isabel Rosenblitt, both statisticians, for their efforts in helping with the effective statistical system behind the original HASA process. Both have since moved to different assignments within HP.

The author would also like to thank Joe Mantz, Reliability Engineer and Senior Member of the Technical Staff at AT&T Wireless; Bobbie Yip, Reliability Engineer at AT&T Wireless; Douglas Ley, Reliability Test Technician at AT&T Wireless; and Mike Brand, Statistician at AT&T Wireless, all in Redmond, WA, for the time they spent reviewing and providing constructive comments on this material. Mike Brand also contributed by writing part of chapter 4.

Foreword

Overstress testing—the process characterized by the application of stresses to a product in excess of its design specification—has become one of the principal methods used by manufacturers of electronic products to improve the field reliability of their products.

There are two principal techniques involved. The first is to generate a failure, find the root cause of the failure, and then eliminate the causes by design improvements. This is generally done during the original design of the product. Next, once a product has been designed and is being produced and sold, the field reliability of the product can be improved by finding and removing the weakest members before they are shipped.

Neither of these ideas is new, except for the fact that overstresses are used. The seemingly minor question of how much overstress to apply and when to apply it to how many of the product has generated countless technical seminars, workshops, and even specialized new divisions of established technical societies such as the IEEE (Institute of Electrical and Electronics Engineers) and IEST (Institute for Environmental Sciences and Technology). As a final measure of the maturity of the discipline, even some of the acronyms are trademarked.

The author of this book has been a principal contributor to this new discipline for most of the 30 years of its existence. He takes us on a detailed tour of the techniques that evolved at Hewlett-Packard. It is rare that one gets to see the internal workings of a great company—particularly in a "know-how" discipline such as this. Each chapter is full of guidelines and "rules of thumb" that are normally unavailable to outsiders. As the author states, the book is written for novices and more experienced engineers who are looking for useful information to help them produce highly reliable products.

It is not particularly the answers given in this book that are the most valuable, but the questions themselves, and the methods used to obtain the answers. One must finally keep in mind the wisdom of Monty Python—"You have to work it out for yourself."

Edmond L. Kyser, Ph.D.
Cisco Systems Inc

Preface

This book has been prepared with both novices and experts in mind. It has been written so that either can find information that will aid them in their quest to produce high reliability products without getting bogged-down in equations. HALT, a process for the ruggedization of pre-production products, and HASS, the production screen for the products once they have been characterized in HALT, are the primary focal points in this book. For those wishing to delve into more advanced topics, three versions of a production audit, HASA, are also included. These may be of interest to the high volume producer or to those who wish to audit their overall production processes rather than to screen all of the products.

HALT, HASS, and HASA all utilize very high rate of change temperature chambers, which are combined with multi-axis pneumatic vibration systems. These are only two of the stresses to which a product will be subjected. Others are also presented and discussed.

I wish to share with you the knowledge that I have acquired in helping employers and clients manufacture products which far exceeded stated reliability expectations as soon as they were released into production. Many who are using these techniques are included in the world's top 50 successful electronic and electromechanical companies. A very important fact that I would like to also share with you is that these techniques are only limited by the imagination of the person wishing to apply them and not by the techniques themselves. In other words, let your imagination run wild. I would also recommend that the reader obtain a copy of the papers listed in the Reference section as reference material.

Over the years, many have applied these techniques with phenomenal success. Some have seen their market share climb from "the dust in the rear of the pack" to be the industry leader in a short time. Other companies have used HALT, HASS, and HASA techniques to increase warranty coverage for their products three times the industry norm, and still reduce their overall warranty expense as well as to dramatically increase their market share while others have used them to assist in the winning of the coveted Baldridge award. These are some the *hidden jewels* that may not be obvious to the novice or to even those experienced in the *traditional* ESS field. Traditional

ESS is a small portion or subset of the overall processes that are presented within these pages and the techniques herein go far beyond them. As you can see from the Reference section at the end of this book, some have been writing about their success while others continue to use these techniques as a way to remain ahead of their competition and do not wish to publish. Their decision, as well as those former clients who wish to remain anonymous and their confidentiality not disclosed, is respected.

My wish is for the reader to use this as a resource book and a place where all that is required to implement a successful accelerated reliability program can be found. I hope that this book will be as helpful to you as the techniques that are presented herein have been over the years to my employers, clients, friends, and me.

All of the references that are used in this book are listed in the references section at the end of this book. Each reference is noted by a^x, and is found by that designator in the Reference section.

ONE

The Importance of High Reliability Products at Market Introduction— How and Why to Do a HALT

> There are truths which are like new lands: the best way to them becomes known only after trying many other ways.
>
> **Deni Didreau**

The proliferation of Highly Accelerated Life Test (HALT) and Highly Accelerated Stress Screen (HASS) has been dramatic during the last few years. HALT has led the way because it yields products that are extremely robust and don't fail in the field. HALT is a direct offspring of the old Environmental Stress Screen (ESS) process and it has taken on different names as it was adopted into different work environments. Some have called it Reliability Enhancement Test (RET); others have called it Stress for life (STRIFE); some have called it Accelerated Reliability Test (ART); yet others have called it Accelerated Stress Test (AST); while some have implemented it as HALT. Some of the rationale for the acronym change is strictly cultural. Management and the technical community felt more comfortable with a name that they already had or were familiar with. The other acronyms mentioned in this chapter may resemble HALT but they may or may not actually be considered true HALT. For example, STRIFE may or may not use a combined thermal and vibration chamber and therefore, combined environments stress would not be done. In this case, STRIFE is not HALT.

This book will focus on HALT, HASS, and Highly Accelerated Stress Audit (HASA) as they have been successfully implemented in various companies throughout the world.

Introduction

Some manufacturers have made considerable strides in improving product quality during the 1990s however, dramatic product reliability improvements have eluded many companies. Often those companies that have found new methods to achieve profound changes in product reliability either publish very little or nothing. Fortunately, some successful companies have been willing to share their results publicly. This chapter addresses the methodology called HALT as used by these companies. The rationale for HALT, its justification to management, and where it will lead your company are all discussed.

An Overview Of HALT

As mentioned previously, HALT is known by many names (STRIFE, RET, ART, and AST to name a few) but for the sake of consistency, we will examine what it is and how it works. HALT is a result of an evolution of product stressing stemming back to the Environmental Stress Screening (ESS) days of the 1960s. Its evolution was a result of the discovery that traditional methods did not cause latent (dormant) defects to become patent (active and detectable). Many of the stressing methodologies of those early days are still in use today with very few, if any, modifications, even though component failure rates have improved by orders of magnitude. This fact alone should cause one to wonder if product burn-in is as effective today with these minuscule component failure rates as it was years ago. (Burn-in, as used here, is defined as a constant level, elevated temperature soak or dwell of the product for a predetermined time period.)

HALT constitutes both singular and multi-faceted stresses that, when applied to a product, uncover defects. These defects are then analyzed, driven to the root cause, and corrective action is implemented. Product robustness is a result of following the HALT process. Under HALT, products that have been designed to meet the benign stresses of an office, could be subjected to stresses that are at least twice as severe as those for which they were originally specified to meet.[3] Every defect that is uncovered during HALT is logged into a database (or spreadsheet) with all of its pertinent information, such as the date of occurrence, description of the failure, its root cause, etc.[1]

The Importance of High Reliability Products at Market Introduction 3

This information is then used to determine a product's robustness when compared to its predecessor. The data can also be used for gauging a product's maturity prior to release into production.

HALT has three distinct phases:

1. Pre-HALT. During this phase the test engineers and designers are preparing to execute the HALT. Test suites (software), fixtures, cables, data collection (and format), as well as resource allocations are items which should be considered. Typically, one or more meetings may be scheduled to discuss the progress and to set a date to begin HALT.

2. HALT. During this phase, the HALT is executed according to the plans formulated during the pre-HALT meeting(s).

3. Post-HALT. A few days after the distribution of the HALT report, the same group that met for the pre-HALT meetings should now reconvene and discuss the issues uncovered during the HALT. Assignments are made so that each issue has an owner. He or she will need to report back to the group later on the root cause and corrective action for each issue so that a follow-up HALT may be scheduled. It is highly recommended that a full or partial HALT be scheduled to verify the corrective action since one issue may have been corrected and a new one created.

Why perform HALT? What are its benefits and shortcomings? Why stress good hardware to such extremes? These are all valid questions. Let's address each of them.

1. *Why perform HALT?* HALT is performed so that a product and the process that produces it are mature at product introduction (minimal, if any, reliability growth remains to be achieved) while design and development time are kept to a minimum. A mature product design means greater customer satisfaction, reduced warranty and service costs, and a competitive edge. This will be discussed in more detail later in this chapter.

2. *What are HALT's benefits and shortcomings?* In addition to increased reliability, HALT has provided cost savings and competitive advantages. The support staff of engineers for the newly introduced product needs only to concentrate on product applications and not the fire-fighting introduction issues which could involve research and development (R&D) production, process, and materials (procurement) engineers. Imagine having a mature product at introduction so that your R&D engineers can get back to their bench to design new products and provide your company with new products. Another benefit of an effective HALT program is the release of products earlier than anticipated.[4, 10]

Those fortunate enough to introduce their product before the competition will usually capture the majority of the market share (assuming, of course, proper advertising and a market for the product both exist). They can usually set the price which others will have to follow. Also, one could be aggressive and set a low price at introduction (with substantial profit, of course). This may force the competition to introduce their product at a lower price than they wanted to (lower or very little profit margin for them) or to even cancel the introduction of their product.[10]

An additional benefit to an early mature product introduction is that in today's market the window of opportunity is usually narrow.[18] If the product introduction is delayed, the window can be missed and the product introduction can be canceled.

The only *shortcoming* is the investment of capital involved with going from the traditional technology to HALT and HASS. These issues are covered in detail in chapter 6.

3. *Why stress good hardware to such extremes?* The product may function perfectly under favorable environmental conditions, but what about all of the other environmental stresses that it may or may not encounter between your shipping dock and the end user such as speed bumps, potholes, extreme temperature dwells, and thermal shock? Temperature change rates under the hood of an automobile or in a jet engine can easily exceed 60°C per minute[9] and don't forget vibration! What does one do now? What about office products that will *never* see these stresses? Well, consider the transportation of a product in a truck or car trunk during a very hot or cold day and then placing it in an office for immediate operation. There is a lot of thermal shock in this simple action.

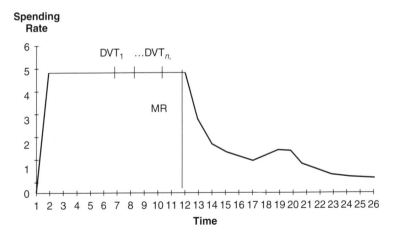

Figure 1.1 Traditional product development.

The Importance of High Reliability Products at Market Introduction 5

The thoughts from the items just described are graphically depicted in Figures 1.1 through 1.3. In Figure 1.1, we see the overall product development cost from a normal product design, qualification, and introduction. When we have the product ready for testing we subject it to some verification testing, correct the problems found, and finally have production release of a product that we think to be defect-free. At Manufacturing Release (MR) the spending rate decreases because the development activity is thought to be complete. (Please note that the *bump* following MR usually is not a single occurrence event and the expense for correcting widespread defects can and usually does exceed the entire product development cost.) This event is usually a design or process problem that went undetected during our product qualification testing or was seen during development but not understood. It will require the efforts of designers and others for correction. Its overall effect, of course, causes delays in other programs (because of resource allocation) as well as customer dissatisfaction and loss of profit.

In Figure 1.2 we began with the same spending rate shown in Figure 1.1. When the product is ready for Design Verification Testing (DVT) as in Figure 1.1, we begin an intense period of HALT. During this time, we have our designers and test engineers, a HALT specialist, failure analist, and any other expert involved so that the end-result of this effort is a mature product. This effort is depicted by the step increase in the spending rate (between time periods 7 and 9) in Figure 1.2. Notice that after MR we don't have the anomalies associated with the product in Figure 1.1. In Figure 1.2 the R&D engineers are released from their design activity with the product and allowed to develop new products, which of course, contributes to the lifeline of the facility. Although not depicted, customer satisfaction can be at an all-time high.

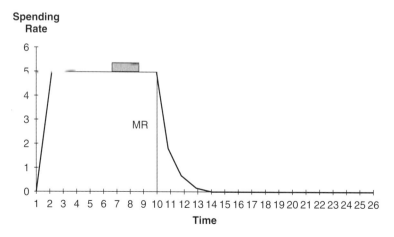

Figure 1.2 Product development with HALT.

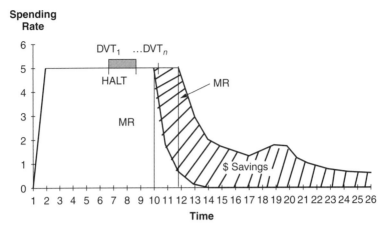

Figure 1.3 An overlay of Figures 1.1 and 1.2.

Notice that when the above figures are combined (as in Figure 1.3) some less apparent events take place. First, MR can and usually does happen earlier although it may not be so for the first couple of HALTs. Second, the product's total cost from inception to obsolescence is far less allowing higher profits throughout the life of the product. Third, economies of scale and other purchasing benefits may be obtainable because of processes that are within control and well defined. All of these events lead to higher customer satisfaction, sales, and profits.

Comparisons of Products with and without HALT

Figures 1.4 and 1.5 help to clarify some of the impacts HALT and HASS have on the product's reliability. HALT impacts every phase of the product while HASS primarily impacts the infant or premature failures. Let's look at Figure 1.4. This graph represents the life cycle of a product developed without the benefits of HALT. The introductory or early failure rate is high and this is followed by a high field failure rate represented by the *flat* portion with aberrations. These are the same aberrations we sometimes read about in the newspaper. They are the product recalls or issues that arise because of out of control production processes, design deficiencies, and a production process which is not monitored with a stress process like HASS (see chapter 2). We have all experienced this phenomenon since we all have returned products for exchange due to premature failure(s). Notice that during this portion of the product's life, additional overhead is provided to support the product in production and in the field. The last phase of the product's life is depicted by the wearout mode.

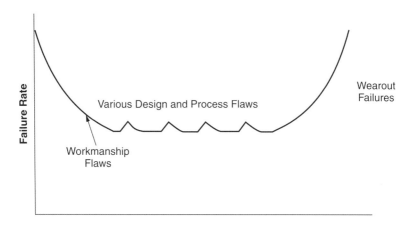

Figure 1.4 Product life without HALT.

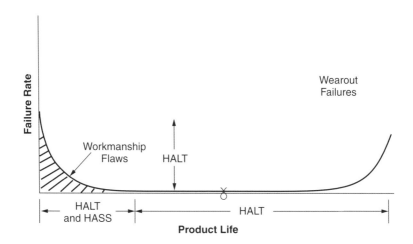

Figure 1.5 Product life with HALT.

Figure 1.5 represents a completely different method of product reliability assessment. As mentioned previously, HALT and HASS are used to greatly enhance product reliability. (Please note that neither axis is drawn to scale but have been exaggerated for this example.) The early failure rate is basically non-existent since we are using HALT and HASS to eliminate them. The area of constant failure is now flat, on the zero-axis, and is much longer than the one shown in Figure 1.4. In other words, the product is introduced as a mature product and failures in the infant and constant failure rate portions of the graph are almost non-existent. Also noteworthy, is that the time to reach the wearout mode is also extended. Most of the HALT fixes can be done at a low to no cost since they are due to incorrect component or material selection and design flaws.

The HALT Process

The best time to perform a HALT is during the earliest product development stages. The underlying assumption is that the product has sufficient test coverage (a goal could be ≥75%) so that the product can be *fully* tested during HALT. HALT is not designed to pass or fail—it is simply done.

A logical question is, how do you do HALT? The stress equipment that is to be used for HALT should incorporate the latest in thermal and vibration technologies, although stresses should not be limited to those two exclusively. (See Figures 5.2 and 5.3 in chapter 5 for examples of these chambers.) Liquid nitrogen (LN_2) is the preferred coolant so that very rapid temperature change rates can be achieved on the product without damaging good hardware. LN_2 also allows for shorter thermal profiles and quicker HALTs. Heating is accomplished through banks of multi-phase electrical heaters and thermally optimized chamber designs that provide product temperature change rates up to, but not limited to, 60°C per minute. These fast temperature change rates are also due to the multiple high velocity fans in the chamber. Vibration uses pneumatically actuated vibrators that provide six degrees of freedom due to their design and mounting configuration. Six degrees of freedom vibration can be best visualized as the three orthogonal vectors (x, y, and z) and three rotational vectors (pitch, yaw, and roll) around each of these orthogonal vectors. This provides non-stationary and non-coherent energy that, in effect, generates an almost random vibration signal.[8] This vibration is analogous to flying in a plane during turbulence where a passenger is not exposed to a single force vector in the perpendicular plane to his seat, but in actuality, the same six degrees or forces of vibration are acting independently to move the body. Stresses (temperature and vibration) will be applied independently and then simultaneously during HALT.

Once the stress equipment is in place, the product is placed in the chamber and connected to test (product monitoring) equipment located external to the chamber. This test equipment is used to provide input to the product under stress and to monitor its output(s). Power must also be applied to the product during HALT; if not, many of the defects will go undetected. Defects that are intermittent, rate sensitive, range sensitive, or dwell time sensitive will all be missed. If the product has on-line diagnostics or if the diagnostics can be downloaded via a data link, a stress that induces a failure can be recorded for subsequent analysis and corrective action. The better the diagnostic coverage, the better the fault isolation that can be accomplished. If the product does not have any diagnostics then some form of exercise for fully monitoring the product for abnormalities needs to be developed so that HALT's value can be fully realized.

It is highly recommended that the HALT engineer meet with the design and test engineers to discuss issues which will need to be addressed before and during HALT. Decisions regarding which stresses need to be applied to the product, what constitutes a failure, and which parameters are to be measured, are some of the issues which require discussion. Consider all possible stresses that can be applied to the product during HALT in addition to temperature and vibration. It is preferred that measurements be made on the critical specifications of the product under stress conditions rather than monitoring go-no-go indicators. The rationale for this is apparent: root cause for each failure must be obtained and if many product specifications are being monitored and data being accumulated, this can be of great interest to the designer. A go-no-go type monitoring is usually not as useful. Also, be sure to discuss product availability for HALT, who will procure the cables to interconnect the product with the test equipment, fixture design and test equipment, etc.

Frequently, failures will occur and HALT will need to be stopped, root cause understood, and corrective action taken. A couple of examples include: a component that is improperly mounted or a poor solder joint that opens during stressing. A recoverable or soft failure, called an operating limit, is not a binary event but has some statistical distribution. This is shown in Figure 1.6 as a normal distribution, which may be the case, or in some other shape (bimodal, skewed left, etc.). The important thing to remember is that this limit will become more clearly defined as more units are stressed. The key is to drive the problem to root cause through failure analysis. Then the corrective action is implemented so that the problem is eliminated. The

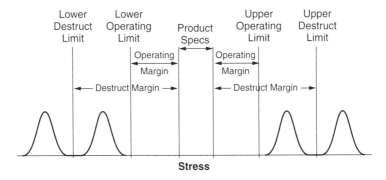

Figure 1.6 HALT margin discovery diagram.

defect is to be readily corrected and the HALT resumed. Don't forget to record the failure(s) as well as the corrective action taken in order to expedite the HALT, as this information may be extremely valuable later to improve design, production specifications, and processes.

Please note that the length of the dwells at each step during stressing should be 10 minutes in duration. If a product diagnostic suite takes longer than 10 minutes, lengthen the dwell accordingly and be consistent (for this particular product HALT, use the same dwell time throughout). The 10-minute dwell provides for product thermal stability and consistency between product HALTs.

A fixture should be made for HALT that allows air to flow freely around all of the product's surfaces. If HALT is to be repeated on many products having the same geometry as the first product, consider designing a longer-life fixture from ULTEM or a similar material. A word of caution; DELRIN is similar to ULTEM but cannot withstand the high temperatures required in HALT. It is not recommended. If the product is to have HALT done once or twice, simple fixturing can be made from channel aluminum or aluminum bars. If a fixture was being designed for a printed circuit board with 6-inch square dimensions, the author suggests that two sections of 12-inch aluminum be cut. Three sections would be used to space the board off the vibration table, with a set of holes for securing the aluminum sections to the table. Short spacers would then fasten the board to the sections by using four or more mounting holes on the board. The fixture should be characterized before securing the HALT boards in it. The characterization will use a nonfunctioning board with accelerometers and thermocouples attached to the board. The goal is linear input to product response.

Cold and Hot Step Stressing

To complete cold and hot step stressing, a product thermocouple (a transducer that converts thermal energy into electrical energy) is attached to a point on the product that "represents" the thermal response of the product. Do not use a high-mass component or one that heats substantially for this purpose. These may be monitored independently. A viable point to attach the thermocouple may be on or near the board's center. Also, be sure that all of the chamber's thermocouples are inside the chamber. Failure to do so may result in product and chamber damage. After performing baseline (20°C, no vibration) diagnostics on the product, the temperature is then lowered by 10°C, the product is tested, and dwelled for 10 minutes. See the step stressing model in Figure 1.7. The process is repeated until the product begins to perform abnormally. Cold temperature step stressing is performed first because it is usually the least destructive of the three generic stresses (cold, hot, and vibration). Don't forget to apply the product specific stresses during each step. At the point of

The Importance of High Reliability Products at Market Introduction 11

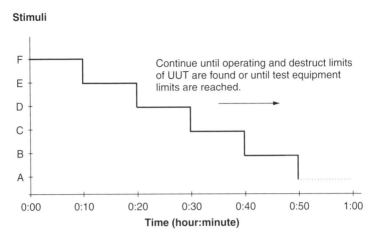

Figure 1.7 HALT step stress approach model.

the product's first abnormality, the temperature is to be recorded and labeled as the Lower Operating Limit (LOL). See Figure 1.6 for an example of LOL. At this point root cause and corrective action should be completed and actions should be recorded for analysis. The process of step stressing continues until the following occurs: the product no longer functions; the product doesn't return to operation when the stress is reduced; or the temperature limit of the chamber is reached. At this point the temperature is recorded and labeled as the Lower Destruct Limit (LDL). The product is then returned to ambient and the defect is corrected and recorded for analysis. The same procedure is then repeated for increasing temperature. The Upper Operating Limit (UOL) and the Upper Destruct Limit (UDL) are recorded using the same criteria mentioned for the LOL and LDL. Once again the product is returned to ambient and corrective action is taken if a failure occurred.

A question may arise: What if the product is too costly to be subjected to the Destruct Level (DL)? It is important to realize that the DL does not imply that the product is totally destroyed. It means that it is inoperative and that some level of repair will be required to restore its full functionality. Record the stress level and observations regarding the product. In the simple case of performing HALT on a notebook computer, the DL could be when the display blanks (at high or low temperatures) and does not recover (even after the stress has been removed). Even though the computer itself may be still functioning (the display doesn't but the PC does), the entire PC system is not performing its intended design function which is to provide information to the user. As a result of thermal step stressing we have determined the product's thermal time to failure weaknesses. With rapid thermal transitions, we will uncover thermal range and rate sensitivities.

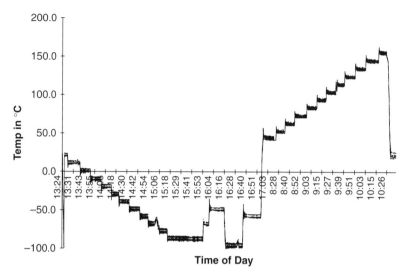

Figure 1.8 An example of temperature step stressing.

We will use these limits for determining the thermal extremes for the rapid thermal transitions and repeated combined stress profile to which the product will be subjected during HALT. An actual temperature step stress plot with the product's temperature is shown in Figure 1.8. The *spikes* shown below are from the air thermocouple in the chamber and not from thermal overshoot on the product. Typically, the air will cause the product temperature to follow to the setpoint. This will cause the air temperature to overshoot but not the product.

Rapid Thermal Transitions

The next step in our temperature evaluation of the product is rapid temperature change rate. For this test we will set the thermal limits to about 5°C or 5% less than the thermal operational limits found during step stressing. Allow the chamber and product to be subjected to the fastest thermal change rates possible while monitoring the product under powered on conditions. Three to five thermal cycles should be applied. Five is preferred. Each cycle should have a 10-minute dwell at each extreme or as long as it takes to achieve product thermal stabilization (if longer than 10 minutes) and execute the test diagnostics. This stress uncovers thermal rate related product weaknesses ($\Delta T/\Delta t$) and thermal range related weaknesses (T_1-T_2) where T is temperature and t is time. Don't forget to apply the product specific stresses throughout this portion of HALT.

Vibration Step Stress

The product is next secured to the vibration table using the simplest fixture possible. This fixture does not need to secure the product or assembly as it is secured in its end-use application (boundary condition). Therefore, additional failure modes may be introduced. In HALT we are not trying to simulate the end-use environment but we are trying to stimulate product weaknesses to surface. As a point of reference, if a defect takes 18 months to surface in the field and one simulates to duplicate the defect, it will take 18 months to do so. If one stimulates the defect to surface, it can be duplicated in far less time using HALT (or other accelerated) techniques. This has been verified many times by the author as well as others who have subjected their product to a HALT. Since the table vibrates in the three orthogonal vectors and three rotational vectors simultaneously, the product can be mounted in any one of the three orientations. Once again, the product's diagnostics are performed at ambient conditions with no stresses applied. Attach the accelerometer (a transducer that converts acceleration into an electrical signal) to the product so that it indicates the point of maximum response. In the case of a printed circuit board that is secured to its fixture by its four corners, this may be the board's geometric center. A rigorous vibration survey is not required but knowledge of the product response should be done to make sure that the product's maximum deflection point is found. This is best done with a non-functional product before the vibration phase of the HALT begins so that a good product is not unexpectedly damaged by a product or fixture resonance. The same step stress procedure that was used in temperature step stressing is used once again. It is recommended that the input or setpoint vibration levels be increased in 2Grms to 3Grms increments beginning with 5Grms and the chamber temperature set to approximately 20°C. The product should remain at each vibration level for 10 minutes, or as long as it takes to execute the diagnostics, whichever is longer. The UOL and UDL vibration limits are also recorded as previously done with temperature. The vibration is then removed when either of these limitations are found. Then the product is returned to ambient conditions and corrective action is taken.

It is highly recommended that following each vibration dwell at and above 20Grms, the vibration level be gradually decreased from 10Grms to 2Grms or 3Grms to uncover anomalies which were not detected during the high level of vibration. This low-level, all-axis vibration is called "tickle" vibration and has been used in many situations. It has been found to be very useful in uncovering defects that would otherwise remain undetected at the higher vibration levels. Tickle vibration is an empirically determined level of vibration and should be sought and recorded for each product during HALT. However, it may not be necessary to perform a tickle vibration determination at each step.

Other HALT Stresses and Special Situations

The step stress procedure is to be included for each additional stress under all HALT conditions. Consider using (in addition to low temperature, high temperature, and vibration) the following: power cycling, voltage margining, frequency margining, and any other stresses that could uncover weaknesses in the product—be creative! Remember that the product is only as robust as its weakest link. Because of this, HALT needs to be performed at the lowest assembly level possible. In most instances, this is at the printed circuit board level. The person performing HALT will need to get clever at times in order to accomplish an effective HALT, because some assemblies can only be separated a very short distance from their input or load before they will no longer function. Some examples include: hard drives and their printed circuit boards or a board that needs to have a set of boards (two or more) in order for it to operate.

Combined Stresses in HALT

The last step in ruggedizing a product during HALT is to combine all of the stresses which were previously used. For temperature, we will use the same thermal profile that was used during the rapid thermal transitions or to within 5°C of the thermal OLs. For vibration, each step will be equal to the step-stress DL divided by five. For example, if we assume that the vibration destruct limit is 50Grms, each thermal cycle will have a vibration dwell of a multiple of 10Grms for 10 minutes. The thermal cycle with 20Grms and higher will be followed by a short tickle vibration. This process repeats through the 4th combined cycle. The vibration level in the last combined environments cycle will have to be set lower than the DL because the product will fail. A product response for this last step of combined environments could be at a level that is slightly less than the vibration OL such as, 3 to 5Grms. In total, there will be five thermal cycles with vibration and other stresses combined. When these stresses are combined, the margins encountered during the step stressing may not be as high as previously found. If the margins were pushed to the Fundamental Limit of the Technology (FLT) during step stressing then not much more can be done to further ruggedize the product except to change the material and therefore the technology. A simple example could be to change from a plastic front panel (which deformed or melted during hot step stressing) to an aluminum or a higher temperature plastic panel. Further investigation will be in order if new failures are uncovered during the combined stress phase of HALT so that the product can be further ruggedized. Once the product vibration response reaches ≥20Grms, tickle vibration should be applied. An alternate method of tickle and high-

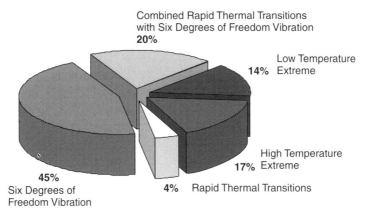

Figure 1.9 HALT defects by environment.

level vibration is to modulate the vibration. In order to modulate the vibration, step from no vibration to just below the OL in discrete steps. Then, decrease the steps back to zero. With this technique in place, not only does the regimen have multi-level vibration levels during the temperature dwells but also during the temperature transitions.

Occasionally, the temptation to skip or even omit some of the stresses in HALT may exist. It is best to always apply each of the generic stresses as discussed before as well as combined thermal and vibration. A study done on defects precipitated and detected in HALT,[19] indicated that each stress, when applied individually, uncovered some defects; when combined stresses were applied, additional defects were uncovered. This study evaluated 47 different products from 33 companies representing 19 industries. The sequence of the stresses applied to the product was the same for every sample from cold step-stress to combined in a clockwise fashion. From this study, the necessity of applying each stress in sequence followed by a combined HALT, is essential for proper evaluation of the product. See Figure 1.9 for an example of HALT defects.

Verification HALT

A verification HALT is one that is used after the original HALT. The primary function of a verification HALT is to make sure that the corrective action that was implemented corrected the original defect and did not introduce any new failure modes. Although this is called a HALT, it may only include the application of one or more stresses. Since a verification HALT may or may not include all stresses, it is not considered a true HALT.

A Perspective on Implementing Corrective Action as a Result of HALT

The HALT process requires that each issue discovered during HALT be investigated to its root cause and have corrective action implemented before production release. This philosophy can be very costly and time consuming. In this section an alternate method is presented which reduces the cost of corrective action while minimizing the risk of not correcting a significant issue. In an experiment at Hewlett-Packard (HP), 86 of 88 identified issues found in HALT eventually occurred in the field. From this, the following question arises: Which issues can we ignore? Obviously, you do not know ahead of time. The methods of this section were adopted in July, 1998 and had been in a trial usage during the preceding 1½ months through six HALTs. Another six HALTs followed shortly after the adoption and the methods continue today.

HALT Limits & Issues

During the HALT process, product design, material, and process limitations may be discovered for each applied stress. Each of these limitations should be investigated to root cause and have a corrective action implemented. A couple of concerns almost always arise. The first is: Which issues should be corrected? The second is: Do we have to correct all of them? The design team management, who were convinced HALT was the correct process for new product evaluations, wanted a process that would provide resource allocation guidelines for the designers and managers alike. The design team and managers were really interested in correcting the issues found in HALT. They knew the value of HALT and were not interested in explaining away a defect since this does nothing to improve the product reliability. They were concerned about the resources needed to address all of the HALT issues.

For this example on how to apply this innovative process, the product had thermal but no vibration specifications published. The product's specifications are shown in Table 1.1.

Table 1.1 Product's thermal specifications.

Hot Temp Spec	Cold Temp Spec
55°C	−40°C

As mentioned before, some knowledge and statistics from previous HALTs have been applied in this chapter. For this process the following assumptions were made:

- A standard temperature deviation is 4°C and a failure can be fully characterized within ±4 σ. This means that if a sample of products all had the same inherent design or process latent flaw, that within ±16°C (±4 × 4°C) all of the same flaws on all of the samples would be detected with thermal stressing if they were to occur. This, of course, assumes that the failures are all thermal sensitive. The 4σ was derived from the author's and two colleagues' experiences in many HALTs and is not based on experimental data.

- A standard vibration deviation is defined as 2Grms and uses the same assumptions as the thermal process previously described, that is ±8Grms (±4 × 2Grms).

Using the Process

With the specifications depicted in Table 1.1 and the assumptions made above, the next issues to address are thermal and vibration limits for the design management team. Please keep in mind that they will continue to design to their normal thermal limits and utilize HALT to help find out where, if any, the product limitations are. For the thermal specifications from Table 1.1, we now need to add 32°C or (2 × 4 × 4°C) to the specification limits in Table 1.1. This yields the results shown in Table 1.2. The 2 is included because this is a two-sided distribution from the mean.

Graphically, these limits are shown in Figure 1.10. With the 4σ (or 8σ since it is two-sided) level applied to the product's limits, one can be assured that no more than ±32ppm (or 0.0032%) defects will escape. This means that 0.032 product defects per one thousand produced will go uncorrected as a result of modifying the HALT process. We can be assured that the product's specifications will be met 99.9968% of the time.

To further clarify this conclusion, all defects uncovered during HALT between -72°C and +87°C will have root cause understood and corrective action implemented. (It may be simpler to round these levels to -70°C and

Table 1.2 Product limits with proposed statistical limits.

Hot Temp Spec	Proposed Hot Spec	Cold Temp Spec	Proposed Cold Spec
55°C	87°C	−40°C	−72°C

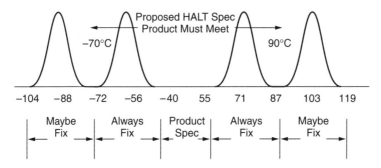

Figure 1.10 Thermal statistics diagram.

+90°C. These could be classified as type A failures. Issues uncovered beyond these limits (classified as type B failures) would have root cause and corrective action understood but may or may not be implemented. The issues with root cause and corrective actions would be taken to the reliability assessment board by the designer and discussed on an individual basis. For each issue, the designer would also need to address the following business concerns:

- How easy is it to correct the issue?
- What are the resources required to correct the issue (people, time, cost, etc.)?
- By implementing the correction, are there any delays in getting the product shipped?
- What are the risks to the business if the correction is not implemented or is delayed?
- Are there any safety concerns with the issue or with its correction?
- Does the issue affect or damage other assemblies?
- What is the return on investment for implementing this correction?

By following this process, the temptation to *explain away* issues is minimized since each issue is reviewed by a review board consisting of the designer, supply line engineering, reliability engineering, and the product manager.

Exceptions must be addressed on an individual basis. An example of one such exception is a product that uses Dynamic Random Access Memory (DRAM) and possibly stops working at -40°C or below. Today, the fundamental limit of the technology for these devices is about -40°C. In other words, there is nothing one can do to make the product better short of redesigning the product to eliminate this weakness. The recommendation, when such circumstances arise, is that one carefully evaluates the flaw, the stress level at which it was detected, and the possible corrective action options. Realize that the HALT process simply accelerates the flaw and it's not a question of whether or not the flaw will occur in the field but a question of when. A flaw precipitated and detected with vibration may or may not be symptomatic of a flawed mechanical design. It could be representative of a flawed process in production. One needs to truly understand the root cause of a flaw before deciding the alternatives, and not the other way around.

Please refer to Figure 1.10 for the next three situations.

A limitation is found at -45°C. What should be done?

At -45°C, a conclusion could be that the limitation was found beyond the product's spec. In this case, why should it be corrected? A look at the statistics indicates that if enough samples were stressed, that failures would occur above and below -45°C. If we assume the mean of the failure mode to be at -45°C then failures could occur between -29°C and -61°C (±16°C of the mean). Clearly, the product from this perspective does not meet its specs. The solution is to correct the problem regardless of resources.

A limitation is found at -60°C. What should be done?

The product could fail between -44°C and -76°C. With only a 4°C (from -40°C) margin the business risk is fairly high. Since this limitation occurred within the proposed HALT spec, the recommendation is to implement corrective action.

A limitation is found at -80°C. What should be done?

The product could fail between -64°C and -96°C. On the surface, this problem should not be addressed. However, a cursory look at the circuitry is in order to see if any component in the design might have been incorrectly chosen. If so, make the change and perform a verification HALT.

Why 4σ and not something less?

The decision on 4σ was made based on the fact that we wanted to be conservative and err on the low side. Table 1.3 depicts three different sigma

Table 1.3 Sigma and deviation impact on HALT thermal limits.

Sigma	Temperature Standard Deviation	HALT Must Fix	Escape (ppm or %)
3	3	<75°C/-60°C	1350ppm or 0.135%
3	3.5	<70°C/-50°C	1350ppm or 0.135%
3.5	4	<85°C/-70°C	230ppm or 0.023%
4	3.5	<85°C/-70°C	32ppm or 0.0032%
4	4	<90°C/-70°C	32ppm or 0.0032%

levels with three different temperature standard deviations based on the -40°C to +55°C specification. The escape rates with 3σ and 3.5σ may be too risky for your business and you may want to decide on 4σ as a better choice. The *HALT Must Fix limits* are rounded values.

Vibration

The vibration limits were established in a similar manner to the thermal limits except that the statistics would be centered about 20Grms. This would mean that all issues encountered during HALT up to 28Grms would be corrected. The 28Grms was determined by adding 4 times 2Grms standard deviation (4σ) to the mean of 20Grms. Issues beyond the 28Grms level would be addressed in the same manner as the thermal issues. Why is 20Grms set as the mean? This level was selected because HASS screens have been found to be very effective with board vibration response levels between 12 and 20Grms.

Please refer to Figure 1.11 for the next situation.

A limitation is found at 28Grms product response. What should be done?

Vibration product limitations also need to be carefully analyzed. A failure at any level up to 28Grms will need to be corrected. Beyond 28Grms, one would turn to the reliability review board. If an integrated circuit (IC) becomes completely detached during vibration, this usually indicates that the technological limit of the device's attachment may have been exceeded. This type of failure is usually reserved for very high product vibration levels, that is, >40Grms for most surface mount technology (SMT) and through-hole technology. At levels higher than 28Grms, an analysis of the failure mode needs to be understood before a decision can be made.

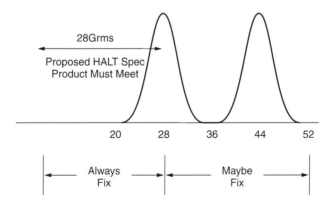

Figure 1.11 Vibration statistics diagram.

Conclusion

Although this process deviates from the *ideal* HALT process (it requires that all issues have corrective actions implemented), business considerations, as well as good common sense, dictate that the deviation is appropriate. It is recommended that you monitor your product's field performance in order to verify if the 4σ statistical level was set too high. Adjust it accordingly from the field results. The thermal and vibration limits used in this example may have to be changed to accommodate your particular product.

HALT Summarized

The following is a summary of what has been presented in this chapter:
Steps for temperature and any bimodal step stressing:

- Place the product to be evaluated in the chamber.
- Attach a thermocouple to the product but not to a heat dissipating or high-mass part.
- At ambient conditions, perform the full product test suite while monitoring the product.
- Decrease the temperature in a 10°C step while dwelling for 10 minutes at each step.
- Perform the full product test suite while monitoring the product.
- Repeat previous two steps until one or more product specs are not being met. This is the product Operational Limit for that stress. Record the out-of-spec parameter (that is, gain, noise, etc.) stress level and determine what can be done to ruggedize product to extend the limit. Correct the deficiency.
- Continue to step stress as before until the product no longer functions or no longer sends data when the stress is reduced or removed. This is the product Destruct Limit for that stress. Record the stress level and product condition and determine what can be done to ruggedize product to extend the limit.
- Return to ambient and repeat the above procedure except increasing the temperature or stress.
- Include product specific stresses at each step.

Steps for rapid thermal change rate:

- Place the product to be evaluated in the chamber.
- Set the chamber thermal setpoints alternately to the operational step stress limits minus 5°C (OL-5°C).
- Perform the full product test suite while monitoring the product.
- Allow the product and chamber to run between the limits as fast as possible.
- Dwell at each temperature extreme for 10 minutes (unless diagnostics take longer to execute).
- Repeat for three to five thermal cycles.
- Return to ambient and repeat this procedure for the next stress.
- Include product specific stresses during the entire profile.

Steps for vibration stresses:

- Secure the product to be evaluated on the vibration table or to its fixture.
- Attach an accelerometer to the product's point of maximum deflection.
- At ambient conditions, perform the full product test suite while monitoring the product.
- Increase the vibration stress in 2 to 3Grms increments while dwelling for 10 minutes at each step.
- Perform the full product test suite while monitoring the product.
- At vibration levels above 20Grms, use tickle vibration to detect defects precipitated at high vibration levels.
- Repeat previous three steps until one or more product specs are not being met. This is the product OL for that stress. Record out-of-spec parameter, (that is, gain, noise, etc.) stress level and determine what can be done to ruggedize product to extend the limit.
- Continue to step stress as before until product no longer functions or no longer sends data when the stress is reduced or removed. This is the product Destruct Limit for that stress. Record the stress level and product condition and determine what can be done to ruggedize product to extend the limit. Correct the deficiency.
- Return to ambient and repeat this procedure for the next stress.
- Include product specific stresses at each step.

Steps for combined stresses:

- Using all of the stresses from the individual product ruggedization process detailed above (step stressing), they are to be combined at slightly less than their respective operating limits and applied to the product for three to five cycles with no failures. A good starting point is to use the same rapid thermal transitions as we used above and combine this with vibration stepped in DL/5 steps for each thermal cycle. Adjust the vibration level during the last combined cycle to something less than the vibration OL. This is to say that during the first thermal cycle the vibration would be held constant at DL/5 and stepped to 2xDL/5 for the second cycle, 3xDL/5 for the third cycle, and so on along with all of the other stresses as well.

- Include product specific stresses during the entire combined profile.

Illustrating the Value of HALT

The following is a list highlighting some successes from published material on HALT and from public presentations. This information is condensed here for brevity. Please obtain a copy of the referenced material for additional information.

1. During HALT on a new airplane radio, Magnavox found many defects at stress levels that exceeded their qualification levels. After careful analysis, they determined that these defects would have caused eventual field failures.[5] Their qualification levels were from +71°C to -54°C and 15Grms using an Electrodynamic (ED) shaker. They actually performed HALT using a new combined stress system, from +160°C to -95°C and 20Grms (six degrees of freedom) on the product. The results uncovered 7 design and process defects in temperature, 16 in vibration and 2 during combined thermal and vibration. Nineteen of these defects were design errors and 6 were workmanship related. HALT was used because of the tight time constraints which they were under and saved an estimated $89,000 (on their first HALT) to say nothing of the time savings and the need for fewer units for HALT testing versus traditional product qualification.

2. Because of the success at one division of HP, another division followed suit by implementing HALT in 1994. The successes (all done in 1993) of the original division included the following on the same product:

    a. A problem which took two and one-half months to uncover using traditional qualification methods, took just seven minutes to uncover by using HALT;

b. A connector breakage problem could not be duplicated using traditional vibration equipment. By using a six degrees of freedom shaker they uncovered it in just two minutes. And lastly;

c. A field problem that took one year to uncover in the field, took one week to uncover through the use of HALT. This information was made public, but unfortunately, was not published.

3. The former CEO of Array Technology stated both verbally and in writing that they have not had any field failures since the inception of HALT and HASS in their processes. He made this statement in writing on July 8, 1993, some months after the paper was written and presented publicly.[4] After Array's results were presented, it was clear that a design that had been ruggedized through the use of the HALT techniques could enjoy defect-free usage by the customers. Some of the defects that were uncovered during HALT and some of the issues that they faced with parts suppliers who provided parts that failed in HALT are discussed in detail in the paper.

4. A manufacturer of medical electronic equipment uncovered a number of defects during HALT that would have led to field defects had they remained uncorrected. During the final stages of the R&D phase for this product, I was hired as a consultant to help them with their HALT program. They had developed their HALT program using a modest product temperature change rate (approximately 30°C per minute), temperature extreme storage, and six degrees of freedom vibration, but had completely missed using other very important stresses which were recommended, such as power cycling, voltage margining, and oscillator frequency margining. Power cycling was implemented and a design flaw was uncovered almost immediately. At low temperature with power cycling, a repetitive failure was noted which would not happen when the temperature was changed just 5°C! This was found to be a defect in the display drive circuitry. One needs to carefully consider all of the potential stresses for each product and realize that what is good for product A may or may not apply to product B. Experienced assistance may be required (in most cases it is) to design and perform the first HALT or so.

5. Ultimate Technology Corporation shared findings on their point-of-sales computers that they subjected to HALT because of very high 46% field failure rates that they were not able to duplicate in-house. These computers are used as cash registers in many businesses as well as part of their inventory control process. After they performed HALT the field failure rate was reduced to about 4%. Their HALT findings included power supply defects and keyboard crystal failures during vibration. All were original equipment manufacturer (OEM) assemblies. Single in line memory

module (SIMM) plating problems (vibration), capacitors (vibration), fan bearings (during combined), alphanumeric displays (hot step stress), LCD (hot and cold step stress), and battery clips (vibration) were the issues found during their HALT. Their progress continues to improve the product reliability.[22]

Some Thoughts Regarding Ruggedizing a Product Prior to HALT

Over the years, some issues recur as different products are evaluated through the HALT process. Since there is commonality between some design weaknesses, a list of the more common issues that may need to be addressed prior to HALT has been included.

1. *Avoid sockets for integrated circuits.* During the early phases of a product's development and even into early production, sockets for integrated circuits may be needed (until the code is stable). It is highly recommended that they be removed as quickly as possible to eliminate a high field-failing component.

2. *Latching of PC assemblies in a card cage or a motherboard.* Experience indicates that when boards are not securely retained by ear-latches, brackets, or positive locking devices, they will become intermittent during testing and in the field.

3. *Securely fasten all bolts and screws.* If possible, use a torque wrench or an adhesive and automate the screw insertion process so that they are properly secured and won't loosen during the product's life.

4. *Keep high-mass components away from the PC board center.* The center of the board will tend to be the point of maximum deflection during vibration and small mass components should be mounted here. Keep the large components along the outer perimeter of the board whenever possible.

5. *Provide high level of PC board testability.* During HALT, the greater the test coverage, the better the defect discovery. The test loop needs to be repetitive.

6. *Place components on single surfaces.* Components which are soldered to a PC board and then attached (via a screw) to a panel or cover to act as a heat sink, usually will fail during vibration. It is recommended that either the heat sink be added as a slip-on type or that the board and heat sink not constrain the component's movement.

7. *Cluster electrolytic capacitors.* Low mass, low voltage electrolytic capacitors, which are vertically mounted, should be physically grouped and then a small amount of material, such as RTV, applied so that they move as a single mass. This tends to eliminate problems of one or more breaking off from the board. It is also recommended that at least one of the capacitors be mounted at 90° to the others so that it will vibrate in a different plane than the others for added rigidity after applying RTV.

8. *Use quality wire.* For internal cable harnesses or interconnections, it is recommended that regulatory approved wire be used. These have a higher temperature rating, more insulation, and are therefore less prone to failure due to fatigue.

9. *Simple brackets breaking.* Sometimes, simple 90° brackets break at their bend. Although it may seem strange, notch each end of the bend with a half-circle to reduce the stress concentration at the bend if breakage occurs during vibration.

Although the listing above is not all-inclusive, it does cover many of the more common issues. Even after correcting for the obvious as early in the design as possible, simply subject the product to HALT and let HALT find the product's weaknesses and then correct the problem(s) using the techniques and process outlined in this chapter.

The Recording of Failures and Corrective Action

As mentioned earlier in this chapter, the recording of the problems that are encountered during HALT or HASS and all of the subsequent steps is extremely important.[1] This information becomes the basis for product decisions.[23] For instance, suppose that an engineer wants to use a particular part in a new design. That engineer could go to the database and perform a keyed search on that part by application or even by failing environment. A manager may use the data to determine what additional resources are needed to keep the project on schedule. See Figure 1.12.

There are six distinct status levels which a problem will be assigned as it progresses from discovery to closure (see Figure 1.12).

Each step is important and must not be omitted. For instance, experience has shown if status 5 is skipped that the problem will eventually resurface on a future product.

Figure 1.13 is a graphical historical compilation of the database. Notice that in order for the product to progress from a lab prototype to a production prototype, one would like to see all of the issues depicted as they are in time

Status	Description
0	Cause unknown
1	Root cause has been isolated
2	Solution has been designed
3	Solution implemented in all protos
4	Solution has been verified
5	Problem and solution have been recorded in a *lessons learned* data base so that is will never recur

Figure 1.12 Device Defect Tracking status description.

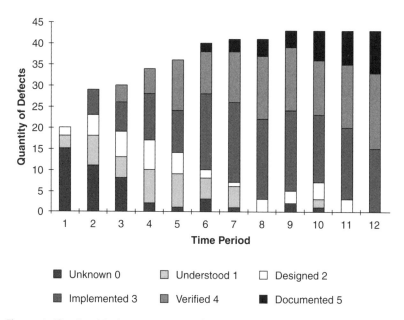

Figure 1.13 Graphical representation of DDT.

interval 12. This may only be possible after management has allocated the necessary resources. This process should not be used as a measurement criteria for different designs or be feared by the designers because defects may be used as a yardstick of comparing designs. Rather, it should be used as a management resource allocation tool only.

Troubleshooting Products Under Stress Conditions

During accelerated testing, products may fail and it becomes incumbent upon the engineer to restore functionality or to improve the margins of the product without (usually) being able to attach test equipment or probes to the product. Let's take a look at some methods of acquiring data on the product under various stress conditions:

1. Failures that occur under vibration may be obvious and require the use of an adhesive, a change in the attachment of the component to the product, or even a change in the component geometry. Regarding the latter, changing from a two-lead vertically mounted crystal to a surface mount device, a low profile component, or one which is mounted parallel to the board could resolve the problem of a crystal being detached during vibration from a board. The attachment of a mini accelerometer to the area of concern may assist in the determination of the course of action that is required by knowing the product response at the flawed site.

2. During cold temperature step stressing, a product begins to fail. Once the failure is isolated to a component, a simple method of attaching a heater to the component can help confirm whether the suspect component is at fault or not. The following steps could be used for fault isolation;

    - Rough up the component surface so that epoxy and one or two power resistors and a thermocouple can be attached.

    - Attach power resistors (1 to 10 Watt) that cover all or most of the surface of the component in question. A higher rated resistor is better than too low a rating.

    - Attach a thermocouple to the component surface along with the power resistors being on top of the thermocouple with epoxy.

    - Connect the power resistors to a power source.

    - Place the assembly back in the chamber but do not turn on the power source at this point.

    - Verify that the failure continues to occur at the same stress level to ensure repeatability.

    - Turn on the power supply so that the component's temperature causes the problem to disappear.

- If the problem disappears, turn the power source off allowing the component to cool and the problem to return. This confirms that the suspected component or its software is at fault.
- Reapply the power source and continue the step stressing.

3. For failures in hot step stressing, consider using a low-pressure (10 to 20 PSI) compressed air attachment with flexible copper tubing (with a piece of heat shrink tubing over the opening to prevent the copper from shorting out component leads) instead of the power resistors as indicated above. This compressed air should be directed over the suspected component so as to maintain its temperature at a temperature lower than the rest of the assembly. Canned coolant can be used instead of compressed air as well as low-pressure gas nitrogen.

Conclusion

As HALT becomes more widely accepted, product reliability will improve and, as consumers, we may be paying less for those goods. This statement is made on the assumption that the cost of the products will dramatically decrease because of the greatly reduced number of engineers (and the associated overhead as will be seen in chapter 6) required with the HALT paradigm. Remember, if a closed loop corrective action program with HALT is implemented, the product should be introduced without defects and in essence, be a mature product. One will not need the large staff to support the product following its introduction and hence, these engineers can be assigned to productive tasks. Since the product is mature at introduction, the manufacturer will have a much lower warranty cost, associated field expenses and less spares, all hopefully resulting in lower costs to consumers.

Predecessors of these enhanced accelerated techniques have been around since 1970. Some companies still use outdated processes (burn-in, traditional ESS, etc.) that were only effective with high failure rate components as well as with poor designs. While utilizing these outdated methods, they discover that they are not effective and some abandon the reliability techniques instead of looking for a better way (HALT) to improve their product reliability. This is unfortunate and will increasingly put these companies at a greater competitive disadvantage.

Two

Highly Accelerated Stress Screen—HASS

It is easy to see, it is hard to foresee.

Benjamin Franklin

Introduction

A thought that comes to mind following the successful introduction of a new product by making it robust and mature through HALT is: What if my production processes or my supplier's processes shift? What will this do to my customer satisfaction? The obvious answers to these questions pose some interesting topics for discussion:

- How quickly do you get warranty reports from your repair organization?
- What do you do with the reports after you receive them?
- Do you know what defect level your customers are willing to tolerate? Will this tolerance change over time?
- When a supplier process shift occurs, does the supplier act as if it is *his* problem or yours?

Let's look at each of these topics. You need to get warranty reports in a timely manner regardless of the shipping volumes, whether they be a low or medium. However, if there is a large delay from the time the field repair is done until the paperwork or reports reach you, consider alternative methods of getting this information. Some ways to accomplish this may be to install a customer hotline staffed with technically competent people, install an automated method of reporting the field failures, or detect these problems before they get to the field. This chapter will address the latter as a monitor for the production processes so that timely product information can be made available.

The ABCs of Building Robust Products

In the previous chapter, we discussed how important it is to have a mature product at market introduction. The product during its design phase needs to be subjected to a design ruggedization process called Highly Accelerated Life Test (HALT). In the HALT process, design and process flaws are uncovered, driven to root cause, and corrective actions are implemented. Each failure, and all subsequent steps, are saved in a database for future use.[1] If a database is not available for saving and accumulating this failure data, use a spreadsheet until one can be procured to meet your needs. This provides a common source of product reliability information from where to draw the necessary information for product decisions. This database, in effect, becomes the criteria for measuring your product reliability improvement. It won't give you a direct Mean Time Between Failure (MTBF) value but it can show reliability growth for your product. The data can be used as a management tool to assess whether a product should progress from stage A to B in its development. The data may also be used to allocate resources in order to incorporate the appropriate corrective steps. All failures are recorded—even if they are one-time occurrences and cannot be duplicated. As time goes on, these issues will reoccur and their incidence of occurrence will increase.

All of the concepts which are presented in this chapter are equally valid for (HASA), which is discussed in chapters 3 and 4.

Once the product is robust (through the application of HALT) the next logical step is to monitor the production processes using a technique called Highly Accelerated Stress Screen (HASS).[4,6] This involves utilizing slightly lower limits than those that were found in HALT to perform the repetitive stress regimen. Since HASS by its name is a screen, *all* of the products are subjected to the stress regimen—none escape.

Production Product Stress Screen—HASS

HASS uses the same stresses as those used during the HALT combined regimen except they are derated because HASS is primarily used to detect process shifts and not design marginality issues. Temperature range reductions also provide the benefit of a reduction in utility costs. A rule-of-thumb on how much to *reduce* the stress levels from the limits of HALT would be:

- Temperature extremes by 15% to 20% of the *operating* limits (OLs);
- Vibration Grms by 50% of the *destruct* limit;
- Component DC voltage levels to product specifications (a 5 Volt IC would use 4.25 V to 5.75 V); and
- Other stress levels would be reduced to within the component's specification, (like the 5 Volt IC specification above).

Note that most of these stress levels could be well above the rated product specifications but are below the limits that will cause product degradation. Typically, the entire stress regimen can be done in about two to three cycles of combined stresses (see Figure 2.3 later in this chapter). During the first cycle (the precipitation screen), the product is stressed beyond its operational limits. In other words, the product's published specifications may not all be met. During the next cycle (the detection screen) the stress levels are reduced to levels at which a good product will fully meet all of its specifications even though these levels are still beyond the product's published specifications. This raises a question: Why stress beyond the OL? During the cycle in which the product is stressed beyond the OL, the product is powered on, fully functional, and may be monitored. It is during this cycle that we actually precipitate the defects that will subsequently be detected during the detection screen. What we have done, in effect, is time compressed our stress screen and therefore, increased our screen throughput. Most traditional production stress screens don't use this technique and rely on a repetitive and a less strenuous regimen to accomplish the same results in more time (see Figure 2.4). Keep in mind that the regimen dwells need to be as long as it takes to execute the product's on-line diagnostics or the product's thermal stabilization. Ten minutes is a good starting point. If the diagnostic execution times are shorter than the dwell time, then many repeats of the diagnostics can be executed during the regimen's dwells. Diagnostic coverage may become an issue for those products that are time consuming to test because a component failure may occur and go undetected if the diagnostic is now testing another product or a different part of the defective product. Later in this chapter are additional detailed discussions on both precipitation and detection screens.

Why Does HASS Work?

Higher stresses of many types produce failure acceleration factors out of proportion to the increased stress. In most cases the acceleration is exponential in nature. Some cases will be discussed later in this chapter. By screening (HASS), we are separating or sorting the flawed assemblies or parts from the good ones. It turns out that the flawed parts or assemblies will have a higher stress concentration at the flawed site than a non-flawed similar site. An example would be a solder joint with poor wetting or a component lead that had been formed with too sharp of a bend or which had been nicked. The stress concentration at these flawed sites is much higher than a normal assembly which is precisely why the flawed sites fail.

Mechanical stress related fatigue damage follows the relationship,

$$D \approx n\sigma^\beta \qquad \text{Eq 2.1}$$

where,
D is the miner's criterion fatigue damage accumulation,
n is the number of cycles of stress,
σ is the mechanical stress in force per unit area dimensions,
β is a material property measured from fatigue testing and ranges between 8 and 12.

The mechanical stress can be caused by thermal expansion, static loading, vibration, or any other stimulus that leads to mechanical stressing.

The graph shown in Figure 2.1 is called an S-N (for stress versus number of cycles to fail) diagram and is derived from tensile fatigue tests on specimens. It illustrates that the relationship between the number of cycles to

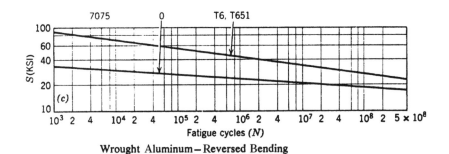

Figure 2.1 SN diagram for T6 wrought aluminum.

fail and tensile stress is exponential in nature and verifies that an equation such as 2.1 exists. β is derived from the slope of the curve. S-N diagrams for other materials are similar, but have a different slope that leads to a different β. The y-axis is labeled as KSI which is pressure in thousands of pounds per square inch. These types of graphs can be found in many mechanical engineering handbooks.

By examining Figure 2.1, we get the following points:

1. At 40 KSI it takes 2,000,000 cycles to fail.
2. At 80 KSI it takes 2,000 cycles to fail.

This information is obtained by following the applied stress (y-axis) and the number of cycles to fail (x-axis).

Hence, an increase of stress by a factor of 2 (40 to 80 KSI) causes a decrease in life by a factor of 1,000 (2,000 to 2,000,000). This acceleration factor is very normal for mechanically induced fatigue. Note that every cycle of stress does non-reversible fatigue damage that is cumulative and cannot be removed.

Parts which fail in service, other than from wear out, usually have some kind of defect which causes an increase in stress concentration. The stress concentrations caused by even a small imperfection may be 2 or 3 times higher than a part without any defects. It can be seen that even a small imperfection can reduce the fatigue life under some conditions by several orders of magnitude.

Vibration

Let us assume that we generate cyclical loading by vibration and that the predominate response which creates the fatigue is at 1,000Hz. (Please refer to Figure 2.1.) The specimen with 40KSI applied would fail in 33 minutes (2,000,000 cycles, 1,000 cycles per second, 60 seconds per minute = 33.33 minutes). The specimen with 80KSI would fail in 1/1,000th the time or .03 minutes, which is only 1.8 seconds!

This example shows the value of using the highest possible vibration level which will, in turn, lead to a higher stress in the assembly. Non-linear effects may invalidate this statement to some degree. A doubling of the spectral density of the input vibration would double the product stress and decrease the required vibration screening time by a factor of 1024 times (2^{10} from σ=2 and β=10). Note that doubling the spectral density would increase the overall vibration level by a factor of 1.414. Using the vibration level factor of 1.414 when calculating the required number of vibration tables can be truly hard to comprehend. For example, if the vibration level were reduced by a factor of 1.414, it would require 1,000 times as many shakers to screen the total production. If one electrodynamic shaker costs $100,000 and could screen the total production at a given vibration level, then a reduction of the vibration level by a factor of 1.414 would increase the shaker purchase cost to $100,000,000! This does not include the cost of test equipment required at each station, personnel, or the new building(s) required to house the equipment and personnel. Hopefully, this would not happen. The power required to run 1,000 shakers of any type is staggering. This may explain why some companies try screening using the "normal" methods and then just give up in frustration because the cost is simply unacceptable. Others try to shorten the screens from an effective time, which may be many hours, and consequently find very little due to the resulting reduced effectiveness. Neither approach leads to a successful competitor in a demanding business climate.

Rate of Change of Temperature

For an example of rate of change of temperature, Steve Smithson reports in the *Proceedings of the Institute of Environmental Sciences,*[14] the number of thermal cycles required to precipitate a delamination problem in a surface mount transistor. There were 400,000 total samples. 100,000 samples were thermally screened in each group run at four different thermal rates ranging from 5°C per minute to 25°C per minute. The number of cycles for the last ones to fail was plotted. The data were replotted on log-log paper. The results are shown in Figure 2.2 as rate of change versus number of cycles to fail. The straight-line fit of the data implies the form of Equation 2.1. It is shown that 400 cycles at 5°C per minute would cause the same delamination as would 4 cycles at 25°C per minute. An extrapolation to 1 cycle at 40°C per minute was made and is shown in the graph. The results quoted are for a specific defect type and may not generally be true for all defect types, but the trend is quite clear.

From Table 2 in Smithson's paper,[14] the rate of change of temperature versus screen time in hours yields these two extremes:

- 5°C per minute required a 440 hour screen time; and
- 40°C per minute required 0.1 hours.

Notice the factor of 4,400 difference! This means that if one thermal chamber running at 40°C per minute can handle a given production (throughput) then it would take 4,400 chambers running at 5°C per minute to handle the same throughput. If a thermal chamber costs $50,000, then the total cost of the chambers would be $220,000,000! Again this estimate does not include the test

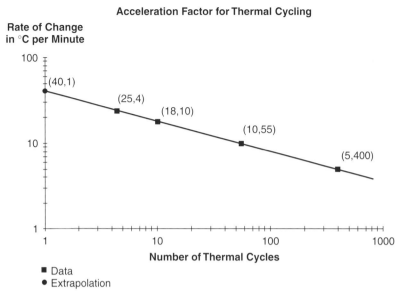

Figure 2.2 Surface mount transistor lifting and tearing up substrate.

equipment, people, or building(s) required to support the activity. A valid cost comparison must take into account the time or number of cycles required and the power and liquid nitrogen (LN_2) used! Also note that less total LN_2 is used cycling at 60°C per minute than at 25°C per minute because of reduced heat transfer through the chamber walls during the shorter time period.

The above example is based on a flaw that is thermal rate sensitive. It is important to remember that not all flaws are thermal rate sensitive. If the flaws are sensitive to the number of cycles and the total thermal range, the broadest possible range will result in the fewest required cycles. Then the screen time will be dictated by the time required to run the cycles and the higher rate will give linear, instead of exponential, acceleration with rate. In this case it is seen that total screen cost is inversely proportional to the thermal ramp rate. With HASS techniques, no dwell time (except for executing the product under evaluation test suite) maybe required because the *product* temperature is controlled during the cycling. HASS screening approaches combine stimuli in one chamber so that the test equipment cost is minimized. There is also a gain in the synergism of combined stresses.

Which Stress Levels are Appropriate?

In HALT, the OLs and DLs are discovered. These levels are not binary events and have some statistical distribution. Initially, this distribution is not known with any accuracy, or perhaps at all, so we must use generous margins between the destruct limits and the HASS levels. Since the various levels will vary with time as the production processes and even the design vary, one should rerun proof of screen on a regular basis. (The process of rerunning proof of screen will be defined later in this chapter.) Additionally, one should rerun HALT or a verification HALT in order to verify the critical OLs and DLs of the product.

The higher the level of the stress used, the faster the defectives will fatigue and break. The non-defectives will accumulate fatigue damage as well, but at a different rate. Equation 2.1 shows that the more highly stressed areas will accumulate fatigue damage much faster than the lower stressed areas. This means that we can fatigue and break the defective areas and only do minute amounts of damage to the defect-free areas. If the design and fabrication methods are robust enough, very high stress levels compared to the field environments may be used to shorten the screens to be very cost-effective.

In HASS we want to use the highest possible stresses in order to compress the screen times. When screen times are compressed less equipment, fewer test technicians, fewer buildings, less power and liquid nitrogen are required.

The first phase of HASS is to combine the stresses found from the limits in HALT. The levels that are to be used for the combined traditional HASS profile should be 80% to 85% of the UOL and LOL for temperature and about 50% of the DL for vibration (the limits were determined during HALT). Please see Figure 2.4. This provides for effective combined precipitation and detection screens. As stated in the previous chapter, the precipitation screen is used to precipitate defects while the detection screen is used to detect those defects. Frequently, when combining the stresses, synergism occurs and the product may fail at a stress level at which it had previously not failed. When this occurs, one or more of the stresses will need to be decreased. The assumption is that the limits encountered in HALT have all been increased to the fundamental limit of the technology. When this occurs, try decreasing the most demanding stress first (such as vibration).

Additionally, some defects uncovered may be non-relevant. A word of caution—do not try to explain a defect away. Root cause needs to be fully understood. Many failures that have been judged non-relevant have caused millions of dollars in warranty expense and customer dissatisfaction. Here are a couple of examples of *non-relevant* defects:

- During the many tests for a plotter, the wheel that moved the paper in the plotting area delaminated or separated the wheel from the grit material. This made the plotter inoperable mechanically. This failure mode was not perceived as important because it had not been documented or understood. After the product was released to production, the plotters began failing at a very high rate in the field due to this problem. It was corrected at a very high cost by the manufacturer.

- During the development of a hard disc drive two failure modes were detected during product qualification testing. These were not perceived as being important since they happened just outside (5°C) the temperature specification of the product. These were not corrected before the product was released to production and caused severe warranty costs for the manufacturer to the point of almost shutting down the business. What had happened was that the failures detected early-on were at the tails of a very wide distribution and as the population of drives increased more units began failing within the product's specifications.

Precipitation and Detection Screens

There are great differences between precipitation and detection screens, yet very little literature can be found regarding these differences. See, however, Hank Caruso's paper titled "Significant Subtleties of Stress Screening"[12] where he differentiates between the *aggravating* environment and the environment in which the defect was detected. Also, see Kam Wong's paper on "A New Environmental Stress Screening Theory for Electronics"[13] wherein he discusses some differences. Let's discuss each screen separately.

Precipitation Screens

A precipitation screen is intended to convert a relevant defect from latent (dormant) to patent (active). Precipitation screens are more stressful than detection screens. An example of a precipitation screen would be a high level vibration that accumulates fatigue damage rapidly, particularly in areas at a relevant flaw, where stress concentrations usually exist. Another example of a precipitation screen would be high rate, broad range thermal cycling which is intended to create low cycle fatigue in the most highly stressed areas, which are usually found at a flaw. Another example of a precipation screen is power on-off switching which is intended to generate electromigration at areas of very high current density and rapid temperature changes which force low cycle fatigue in areas of high stress. (Both functions generally occur near a flaw.)

In HASS, use the highest possible stresses which will leave non-defective hardware with a comfortable margin of fatigue life above that damage which would be done by remaining screens, the shipping, and in-use environments. This approach really demands the application of HALT techniques and design ruggedization, in order to be able to rapidly and effectively precipitate flaws. Without using these techniques, the application of HASS is usually not possible due to product design fragility levels existing below the applied stress level. This leads to the failure of non-defective items. Non-defective, as used here, means those that would survive a normal field environment if stress screening were not done.

Precipitation screens (as shown in Figure 2.3) may be run above an UOL or below a LOL where the system cannot perform normally and therefore cannot be expected to meet specifications during stimulation.

Detection Screens

Detection screens are less stressful than precipitation screens and are aimed at making the patent defects detectable. Temperature slew rates will be lower and the vibration levels may be modulated or at 50% of the destruct limit for some or all of the profile. The reason for changing temperature slew rates and modulating the vibration is that some failures are rate dependent and certain vibration failures are only detectable at low vibration levels and not detectable at high vibration levels. It has been found that many patent defects are not observable under full screening levels of excitation even when the screen is within the operational limits of the equipment. *Tickle* and/or modulated vibration and certain temperatures or temperature rates are required in order for the failures to be observed. *Tickle* means a low-level six degrees of freedom vibration. A suitable *tickle* level seems to be flaw type and product dependent and so some experimental work is required to determine this level for each product. One technique that has been found to be very effective is to use six degrees of freedom vibration for screening, reducing the vibration level slowly at the end of the precipitation screen so that the correct *tickle* level is passed through slowly, providing time to detect an intermittent. For example, on several products that have cracked plated through-hole solder joints, these defects could only be detected by reducing the temperature to the lower operating limit of the product under test and exciting it at a very low level, say 3 or 4Grms. Another plated through hole problem was only detectable under these conditions (which caused a duty cycle of 97% open circuit) and could not be found at all under any conditions of temperature, orientation or vibratory excitation using a single axis shaker.[8]

A detection screen on a computer system (which may take a long time to execute) might be at a very slow temperature cycle (perhaps at <10°C per minute) between the lower and upper operating limits while exciting with low level six degrees of freedom vibration, and simultaneously running the diagnostics. This would probably be a very poor precipitation screen, but could be a very strong detection screen.

Detection screens should be used on equipment returned from the field or production labeled as defective or No Trouble Found (NTF), since we must assume that a patent defect is present or the equipment would not have been returned. It is worth noting that NTFs are frequently returned from the field for various reasons but the main motivator for the field repair staff is created by the customer's *get it running ASAP!* Field repair staff are inclined to replace whole sets of boards or boxes, when maybe only one of the set truly has a problem. If the detection screens do not suffice, then a precipitation screen followed by a detection screen would be in order.

In the case of field returns, it is prudent to *simulate* the field conditions under which the failure occurred. These conditions might include temperature, vibration, voltage, frequency, and any other relevant conditions. The military, airlines, auto manufacturers, and others, could benefit in following this course of action because NTFs account for at least 30% to 50% of field returns. *Stimulation* is not necessarily called for in this case, as *simulation* and detection screens are probably the more effective approaches for field returns.

A Comment on HASS Profiles or Screens

Consider the profile shown in Figure 2.3. This may not be what we are accustomed to using. It can be as effective as the profile in Figure 2.4 and can be

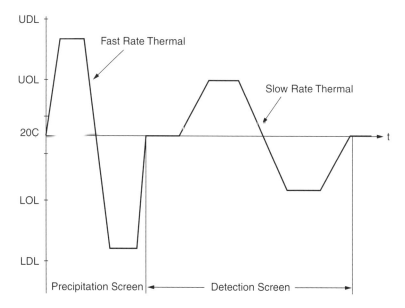

Figure 2.3 An ideal thermal profile.

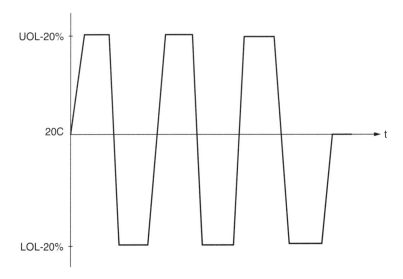

Figure 2.4 A traditional thermal profile.

run in less time. If throughput becomes an issue, consider using the profile in Figure 2.3. Remember, when changing the profile dwell length (in this case, wider temperature excursions), or higher or lower stress levels, a proof of screen must be done before implementing the new profile. Most people are accustomed to the repeated thermal profiles as shown in Figure 2.4. This type of profile provides excellent coverage for defects that can be encountered during HASS. Due to its design and duration, both precipitation and detection screens are one in and the same. This thermal profile is typically combined with vibration and all of the other stresses required for an effective HASS. Although 3 cycles are depicted, the profile can have more or less cycles.

Proof of Screen

Proof of screen consists of three elements. First is fixture characterization, followed by defect detection, and lastly, the proof of life determination. A discussion on each element follows.

Fixture Characterization

The product fixture for HASS may need to be more robust than that which is used in HALT because of the number of products that will be stressed over the fixture's lifetime. The product fixture needs to be lightweight, rigid, have a low thermal mass, and be reinforced to withstand the fatigue damage caused by prolonged stressing. A lightweight and rigid fixture maximizes the transfer of stresses to the product. The product response is important. Vibration or temperature setpoints from the chamber controller are not as important, here. Stated in other words, a fixture can be designed to attenuate most, if not all, of the input vibration or temperature. It may be necessary to reinforce the fixture so that it can withstand the long-term effects of vibration.

The fixture will need to be characterized for both thermal and vibration energy transmissibility. This is best done with non-functioning units, that did not survive HALT or even early prototypes. Keep in mind that HASS fixture design for six degrees of freedom vibration is not the same as those used in electrodynamic systems—they're much lighter and require far less material. A material that has been used successfully in the HASS environment is Delrin™. Delrin™ is easy to machine, fairly inexpensive, and can withstand the daily repeated stress demands. ULTEM™ is another material capable of withstanding a wider temperature range than Delrin™ but it costs more and is therefore better suited for HALT.

Before designing the HASS fixture, consult with manufacturing on their requirements since a fixture to hold 100 products per day could be quite different than one for 1,000 products per day. Another very important consideration when designing the HASS fixture is that the product must be held in the same boundary condition that would exist in its end use environment. In other words, if a product is to be subjected to HASS and is normally held in its end use environment along its two long axis, it must be likewise held in the HASS fixture. In HASS we are trying to simulate. This is not the same philosophy as one would use in HALT which is to stimulate. As part of the fixture design effort, also add holes or openings in the fixture to aid the airflow within the chamber and allow unrestricted flow in the fixture. This also reduces areas of dead airflow. Once the fixture for HASS has been designed and fabricated, it will need to be characterized with non-functioning products. Populate the fixture with as many of these products as practical. For example, if the fixture can hold five products, then three non-functioning products would suffice. Attach a thermocouple to each product and run the same rapid thermal profile as used in HALT while observing the thermal response. This process is used to characterize the fixture and is not the HASS profile. With the exception of a time delay, each product's response should approximate the HALT results. It may be necessary to adjust the air ducting to optimize the product's thermal responses. In addition, air distribution headers may need to be made from sheet metal. Once the thermal optimization has been accomplished, an accelerometer is to be applied to each product and each product's response should be measured. The setpoint should be set so that between 10Grms to 14Grms is measured on each product. If resonances exist, try dampening the material or re-clamping. The end result should be that each product has been subjected to the aforementioned vibration levels. The non-functioning products may now be removed from the fixture. The above process is to be repeated for each fixture position within the chamber. If four or five card fixtures are used simultaneously, then the process should be repeated four or five times; once for each fixture.

During the fixture characterization each of the non-functioning products should be monitored with a thermocouple for thermal response. The thermocouple can be attached to the center of the product, if practical. Eventually, accelerometers should be glued to each product to measure the vibration response as well. The purpose of this, in addition to measuring the product's thermal response, is to adjust the air ducts and fixture so that approximately equal temperatures are experienced on each product. A good rule-of-thumb is to set the temperature to 10°C below setpoint, raising it to setpoint within a few seconds. Overshoot should be minimized. In addition to the thermocouples for each product, it is recommended that a single thermocouple be attached to an area that is representative of the overall product thermal response. This thermocouple point is called a *pseudo load*. Its purpose is to ensure that a thermocouple doesn't get attached to the shippable product and possibly cause cosmetic damage. The location of the pseudo load may not be easy to determine and some trial and error will be in order. Remember that it's not necessary to use a pseudo load if it's possible to use Kapton tape to secure the thermocouple to the product.

The HASS Profile

The selection of the stress limits for the HASS profile is unfortunately not done with the use of a formula. A good starting point would be to use 80% of the HALT thermal *operating* limits and 50% of the vibration *destruct* limit. This would provide the HASS profile that is depicted in Figure 2.4. If during the life of the proof of screen, defects at the thermal extremes are noted, the thermal limit(s) should be reduced until they disappear. In general, a 100°C thermal range and about a 10 to 15Grms product response during HASS can be used in an "effective" screen. These levels can be adjusted according to individual needs. For instance, even though the product may operate throughout a thermal range that is greater than 100°C, it is not necessary to run the chamber to extremes greater than 100°C. No new defects would be detected and only the excessive cost of LN_2, electricity, and time would be incurred.

Defect Detection

The next step in our HASS development is to seed defects and subject them to the combined HASS. Seeding can be extremely difficult, if not impossible, depending on the product technology. For Surface Mount Technology (SMT), it is best to obtain boards that are functional but are classified as No Defect Found NDFs or NTFs. It is also beneficial to try to find a place in the assembly process where products are failed, then retested and passed as NTFs. Typically, these assemblies can have a high fallout rate and are referred to as *seeded samples*. These can then be subjected to the combined HASS and the results can be recorded. They will prove that the HASS combined profile is effective for detecting the population of failure types for which the regimen was designed. Record the number of stress cycles that it takes to uncover the defects from each. As in any effective reliability improvement program, the failure data should be recorded, analyzed, have the root cause determined, and corrective action taken. In the event that the *seeded* samples do not fail, investigate if the defects are located within a signal path that is being tested and then increase the stress slightly to see if the defect is detected. In the ideal situation, all of the defects would fail.

Life Determination in Proof of Screen

Once the stress regimen uncovers the seeded defects, then proceed to the next phase of the proof of screen (POS). Do not skip this step or the seeded sample testing because both are critical. At least six unstressed units or as many units as there are fixture positions, are to be used for this portion of the POS. For this phase of HASS development, the six units should be subjected to the repeated regimen for a *minimum* of 10 times what the normal stress regimen will be. It is recommended that these repeats be in the order of 30 to 50 times. In other words, if the normal HASS profile is three cycles, run the POS for 90 to 150 combined cycles while monitoring the products. Having run all six units as well as the seeded samples (from the previous step) through this combined HASS stress regimen you should uncover two very critical data points:

1. From the first part of the test, ensure that the regimen is robust enough to detect the defects that would normally go through the production process(es) without detection and eventual shipment to customers (they should all fail); and

2. From the second part of the test, ensure that the regimen does not remove an appreciable amount of the product's useful life, (none of the products should fail).

Both of these steps are absolutely essential for a successful HASS program. Now, consider the possibility that the product fails during the normal production HASS and needs repairing, but only 10 passes through the stress regimen had been done as the POS. If the repair of the HASS-stressed product was not successful, then the product may need to be subjected to the HASS profile two or three times (which is possible with multiple failures)

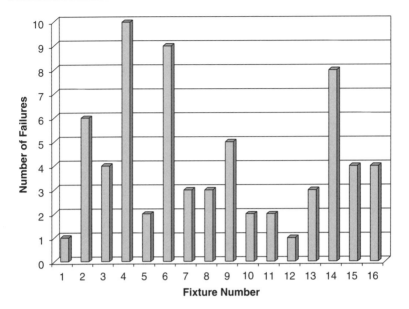

Figure 2.5 Failures by position number for 7,399 PC assemblies.

until the product finally passes. Running the POS only 10 times might not provide an adequate confidence in the *POS margin*. In other words, if an assumption was made that all of the units that were used to validate the POS ceased to function after 10 passes through the regimen, then 10 passes become the "product life." Submitting a defective product through the regimen three times until it is repaired would consume 30% (10% per pass) of the product's "life", which may not be an adequate margin for most products. Based on the same assumptions, running the POS through 30 to 50 cycles would mean a 3.3% and 2% removal of useful life for each HASS submittal respectively.

A word of caution about POS and systems that can and will accommodate multiple products simultaneously. A POS must be performed for each product-position within the chamber.[15] In other words, if the normal HASS will screen 10 products simultaneously, then each of the 10 positions must have a POS done for it. They can all be done at the same time if the monitoring equipment is capable of handling the units under evaluation simultaneously (in order to save utilities, costs, and time). The reason for doing a POS for each position is the variability in both vibration and temperature in the chamber in each position. This variability may be due to the fixture, cables, and any other impediments within the chamber workspace.[15] The POS can be done with all products fixtured at the same time.

In order to clarify this even further, the author did a case study which is detailed here.[15] There was a concern that there may be *hot spots* in the chamber because of the variation in failures from fixture position to fixture posi-

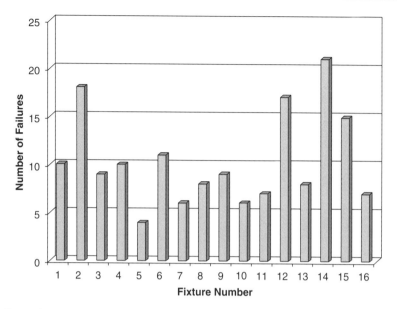

Figure 2.6 Failures by position number for 14K PC assemblies.

tion (see Figure 2.5). There were eight fixture positions each capable of holding two boards simultaneously, back-to-back. Up to this point, a total of 7,399 PC assemblies had been processed through the chamber in an audit mode (HASA). This was discussed with the statistician and it was concluded that as more test samples were processed, the variability would be less. Notice, not only is there variation from fixture to fixture but even from board to board within the same fixture. Please refer to Figure 2.5 for an example.

The graph was then plotted after 14,000 boards had been tested and those results are shown in Figure 2.6. Notice that the y-axis is rescaled and that there is less variability when compared to Figure 2.5. As a side note, every board that had failed during the stressing (in either situation) was always resubjected to the stressing in a different fixture position number, and without exception it always failed again. These findings testify that stress variability within the chamber should not be a deterrent to performing HASS or HASA *if* a POS has been done for every fixture position.

Screen Tuning

After the stress regimen has been in use for some time and the failures have been recorded by the stress cycle in which the failure was detected, the process of screen tuning may begin. Screen tuning is used to refine the stress regimen and optimize it for higher throughput. During the period prior to

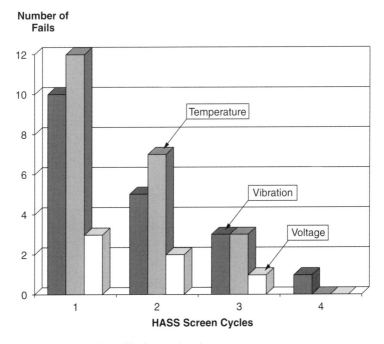

Figure 2.7 Data gathered before tuning the screen.

beginning the screen tuning, the distribution of failures by passes through the regimen might look something like Figure 2.7. The objective of tuning the screen is to cause all of the failures to occur in the first cycle, like Figure 2.8. From the results of Figure 2.7, one would have to increase vibration, temperature, and voltage. Consider, as well, the failures that may be happening on the production floor at system startup as well as in the field and ask yourself if the screen can be optimized to detect these failures as well. Since the regimen has now been changed, POS must be repeated for each product position in the chamber. In today's environment, the opportunity to reduce the number of combined stress cycles or to perform screen tuning is rarely achievable. Regardless, one should attempt to reduce the number of cycles in the stress regimen to increase throughput and reduce the cost per board or unit.

Cables for HASS

An important consideration is the effect that the stresses will have on the signal interconnect system, (the cables and connectors which will interconnect the product under stress to its monitoring equipment external to the chamber). Since the cables can be expensive to replace, consider using replaceable

Highly Accelerated Stress Screen—HASS

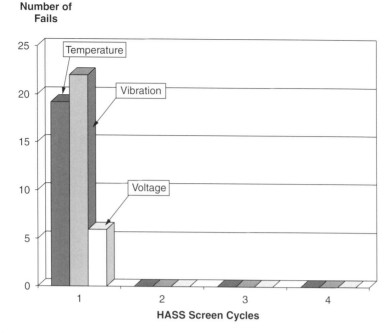

Figure 2.8 The screen after performing screen tuning.

connectors with service loops in the chamber end so that the entire cable doesn't have to be replaced when it becomes fatigued. Closely monitor and record the number of system defects over time. A second method of extending the life of the main cables is to insert short adapter cables that interconnect the product under evaluation with the main cables. When this adapter becomes fatigued, it can simply be replaced with a new one. In addition to the cabling, any boards or assemblies that are part of the test setup may wear out as well. Information on predicting when to replace connectors is provided in chapter 3.

The last issue that one needs to consider is the cable sheathing material. The most commonly used cable sheathing material is Polyvinyl Chloride (PVC). It meets all of the HASS and HASA requirements except one—it produces chlorine when heated above 90°C. This toxic gas can be fatal and it is recommended that alternate high temperature cables be used, such as Teflon.

HASS Summarized

The following list details the steps for HASS.

- Perform HALT (product ruggedization).

Screen design

- Using the limits from HALT and Figure 2.4 (traditional profile) do the following:
  - Derate the rapid thermal limits by 15% to 20% of the OLs; and
  - Derate vibration by 50% of the destruct limit; *or*
  - Select the ideal profile (Figure 2.3).
- Apply all other relevant stress limits according to component specifications.

Fixture design

- Design with product boundary condition as one of the criteria.
- Consider manufacturing requirements (their throughput).
- Design fixture so it is rigid, lightweight, and thermally translucent.
- Consider the air that must flow through the fixture provide openings as needed.

Fixture characterization

- Insert non-functioning products in the fixture.
- Apply the combined profile while monitoring (thermal and vibration) the products in the fixture.
- Adjust air discharge ducts for equivalent thermal profiling on all products.
- Apply dampening material to the fixture if resonances are detected.
- If above two adjustments don't resolve *hot spots,* fixture may need redesigning.

Proof of screen

- Apply the normal regimen with products that have been seeded with defects. All defects must be discovered. If not, evaluate seeded units and repeat regimen.
- Subject a new (unstressed) product for each fixture-position to at *least* 10 times (recommendation is 30 to 50 times) the normal regimen while fully monitoring the product.

- If product does not survive POS, decrease most severe stress slightly until all products survive.
- HASS regimen is ready to be used in production.

Screen tuning

- Run HASS for some time while gathering data to confirm at which combined stress cycle number that a failure occurs.
- Consider which stress(es) need to be increased in order to cause all of the failures to occur in one combined stress cycle. Also consider field start-up failures as well as in-house system failures, if any.
- Re-run POS.
- Implement the stress regimen.

Some HASS Successes

1. Sequent, (now IBM in Beaverton, OR) a computer manufacturer, has been performing HASS since 1994 and has uncovered many issues related to their product assembly process and components.[6] Their HASS process consists of temperature (75°C to -40°C), vibration (modulated from 8 to 20Grms) and power cycling (5 Volts) while performing low-level diagnostics. The ratio of component defects to assembly defects was running between 40 percent and 60 percent. Their boards are predominately fine pitch SMT (approximately 2500, 20 mil joints), with passive components on the backside of the nine-layer 12"×14" board. Approximately 3.5 pounds of heat sink was used on one of their boards and they were able to achieve a 45°C per minute product temperature change rate. Their boards were all ruggedized through the HALT process prior to HASS.

2. Northern Telecom, a major manufacturer of communication equipment, decided to compare HASS and burn-in on power supplies after they had performed HALT.[20] They decided that they would perform burn-in on half of their products and an Environmental Stress Screen (ESS or HASS) on the remaining products. Of the 153 total defects uncovered, 114 were found in HASS. There were defects that were not detected in burn-in that were seen in HASS. The burn-in took 8 hours versus 2.5 hours for HASS.

3. Defects uncovered in Array Technology's disk array products during stressing are well documented.[4] Not only were defects found during

HALT, but also during HASS they found problems with components after the product had been released to production. Their dealings with their suppliers and a good lesson learned is well documented.

4. A different type of HASS success was shared by Javier Oliveros.[7] Although his product failure information was not divulged because of U.S. government restrictions, he stated that the six-degree of freedom system that he uses was more cost effective than other shakers because:

   a. the product did not require fixture axis reorientation in order to vibrate the different axis (their shaker actually produces the three orthogonal vectors and three rotational vectors simultaneously); and,

   b. the cost of this vibration system was "substantially lower than the tri-axial ED shaker machine." The reduction in vibration time with their new system was from thirty minutes down to two minutes! Additional benefits included reductions in maintenance as well as utility costs. By the way, Javier was able to use the six-degree of freedom system and with the appropriate fixture-dampening material, he was able to model the NAVMAT vibration profile!

A Word of Caution

A word of caution is in order once HASS has been implemented and defects begin to happen. If HASS is finding the occasional defect there shouldn't be a problem but once many defects are found the situation may be quite different. What happens is that someone in production (usually) will have assumed that the reason for all of these defects is due to the fact that the HASS stress levels are too high and therefore, damaging good hardware. Nothing could be farther from the truth if a proper completion of the life portion of the (POS) was done. If this was run for the 30 to 50 times the normal HASS profile, the normal HASS profile in production is doing exactly what it is suppose to do—find defects. These defects could be caused by many variables but HASS is not one of them. The potential sources of the problem could lie in the attachment process (if it's a board), a particular component (if the supplier changed a part without your knowledge or approval), a test change in how it is done or an inadvertent change in specifications, or a software problem? Each will have to be investigated and eliminated. Once again, be careful so that you don't compromise the value of HASS when it is actually doing what it was designed to do.

Conclusion

The assumptions of this chapter were written around a complete stress screen, that is, 100% of the units produced are stressed. If no out-going defects can be tolerated or if you must adhere to contractual agreements, then you must remain with a stress screen (HASS) which has been shown to be an effective way of sorting the defective population from the good one. Papers that discuss HASS are available and the examples mentioned in the References section are but a few of the papers that have been published.

A clear understanding of the difference between precipitation and detection screens will allow one to develop screens that take advantage of this difference. Precipitation screens can be run beyond the operational limits of the equipment (if proven to be safe in terms of fatigue damage) in order to accelerate the process and then followed by detection screens that are used to detect the precipitated defects. Two types of HASS screens are available. The arrangement just mentioned is the ideal profile but most people feel more comfortable with the typical profile.

As production processes improve and defect levels decrease, it is appealing to consider using a stress audit (HASA). This, of course, is yet another step in the stressing process. It is recommended proceeding to HASA only if the production (and supplier) processes are consistently within statistical control and few defects are anticipated. The benefits of auditing are tremendous but the risks must also be considered. Stress audits will be discussed in the next two chapters.

Three

Beyond the Paradigm of Environmental Stress Screening—Using HASA

> The ultimate measure of a man is not where he stands in moments of comfort and convenience, but where he stands at times of challenge and controversy.
>
> Martin Luther King Jr.

Introduction

Once HALT has been performed on a product during its developmental phase, the next step is to implement stress screening of the product in production. As mentioned previously, this stress screening is called HASS, or Highly Accelerated Stress Screen.[4] Some assumptions are made when HASS is implemented. It is assumed that the production (internal and supplier) process may not be within statistical control. One may have an underlying objective to reduce the screen to an audit in time. When the processes have achieved statistical control, an audit may be in order. This will increase profit margins because less stress equipment, personnel, utilities, and product test equipment are required. The stress methodology to accomplish this is called HASA, or Highly Accelerated Stress Audit. This chapter addresses a process that was first successfully implemented by the author at Hewlett-Packard

(HP), Vancouver Division, in 1989 on the DeskJet family of printers. The data included in this chapter represents the HASA process with some degree of confidentiality considered. However, the conclusions and business decisions that were made at HP will be discussed in detail. The production processes detailed in this chapter were, over time, improved to the point where a refinement in the statistics was necessary. This refinement and others are discussed in detail in chapter 4.

Background

After attending a seminar on HALT and HASS in 1989, a reliability engineer returned to his job at HP wondering how he could apply all of the new information that he had learned. At the time, HP was building a moderately high volume, thermal ink jet printer (DeskJet Plus) and knew that these production volumes would soon be dwarfed by its successors. But, HP didn't even imagine by how much. One of the biggest concerns that divisional management and reliability engineering had was the adverse effects that a major reliability issue would have on their loyal customers and their image as a manufacturer of high quality, low cost printers. A recent supplier integrated circuit (IC) design change (prior to the implementation of HASA), which was not approved by HP at the time, was discovered in production by chance. This caused some reflection. If this were to happen again, HP would not have the ability to detect it in-house before it would be shipped to customers. The supplier IC design change situation could have been disastrous.

Months were spent prior to attending a HALT and HASS seminar (by others as well) trying to duplicate or mimic in-house the failures which customers were experiencing with their printers. This became an effort of "wheel-spinning" since all of the products had been qualified by using traditional environmental systems and we were trying to uncover those field defects using the same equipment. In an effort to improve the system, an employees were sent to a seminar. Many concepts from the HALT seminar were already being used at the facility but some were not. One engineer had the responsibility of improving the robustness of products during the design phase, while another focused on the production side. In other words, the two were not operating as independent engineers, but were working together so that the information gleaned from the evaluation of the engineering units could be directly applied to what was to be done for production. It is important to add that the seminar was one of the many courses and seminars which the HP engineers attended but it was the only one with "new" information. Most of the other seminars they attended clearly did not offer anything that would be of help to them.

As was mentioned, HP was a moderate volume manufacturer at the time but all indicators pointed to the fact that they were clearly underestimating the potential for thermal ink jet technology. In a short period, they discovered that this was the case. Other formidable challenges were soon upon them. The need to monitor their processes in-house (both theirs and their supplier's) became extremely critical. They needed a *real-time* system which could provide data to reliability, production, and process engineers. They also needed an equally high degree of confidence that when a problem was uncovered it was real. Discussions of these concepts will be presented later in this chapter.

Recalling that auditing was the goal in the HALT and HASS processes, one of the reliability engineers embarked on a project that had auditing as one of its primary focuses. Statistics would be an integral part of this system. Very little was known by the reliability engineer about these statistics but he was soon taught by three excellent statisticians. Many at the facility felt that the production processes were under fairly good statistical control but there was opportunity for certain processes to go out of control and possibly even unnoticed for long periods of time. This, of course, could be disastrous. The objectives that were selected for this monitoring process (HASA) were:

- To detect any sudden increase in failures reflected in the outgoing product stress failure rate that could be detrimental to the business. This detection was to be done within eight hours of its onset. In other words, they assumed a baseline failure rate that was both acceptable to their customers and to their business. If the failure rate stayed at that level or below, they were satisfied. If the failure rate took a dramatic step upwards, they wanted to be notified immediately so that the corrective action team could be activated.

- To provide continuous feedback to their engineers regarding all defect types for future product improvements. The issues uncovered were not of the type that could cause the failure rate to suddenly increase but were of the type where an occasional component would fail and its root cause and failure analysis would be sent to the engineers.

- This process had to be totally transparent to production and could not increase the build cycle times.

Each of these three objectives had their own sub-objectives. For instance, under the first objective, it was decided to assume some risk in allowing defective products to ship but the potential for exposure was extremely low (<10%). Likewise, they were also learning about HASA and were not willing to assume any risks that would cause a sudden increase in the outgoing failure rate to go undetected. In other words, the risk that was assumed in allowing a predetermined number of defects to be shipped would not place the business in any compromising position and furthermore, the increase in defects would not adversely affect the product's baseline failure rate.

Statistical Process Overview

The statisticians were heavily relied upon for helping to find a statistical system that would work short-term. As the processes improved and became more sophisticated, the statisticians were relied upon to modify the statistical system and help with data analysis. The statistical system for this type of process needed the following characteristics:

- Sample size of the units to be stressed must be immune to the rapid increases in production build rates. With large increases in production ramp rates, they did not want to have to constantly recalculate the sample size. In other words, the sample size should not be a fixed percentage of the units built.
- It must be easily adapted to possible changing risks. If they wanted to decrease the risk exposure by either HP or to their customers, it would be a simple matter of changing a value that would affect the sample size.
- It must be easily changed when outgoing or field failure rates changed.
- The system must allow for production process improvements when the anticipated number of failures was less than five during stressing. The original concept assumed at least five failures and they would, over a period of three years, improve the production processes to this point (see also chapter 4).

The last two characteristics appear to be the same but really are not. The second to the last characteristic relates to variations in the outgoing failure rate from processes that are in control but not to the point where five or less failures are anticipated. This was the case for the first two years of stressing the DeskJet family of printed circuit boards. The last characteristic takes the entire statistical system one step further when five or less failures are anticipated within the sampling period. When this is the case, one must use the exact binomial distribution rather than the estimation. One could ask the following question: Why then, use the imprecise equation? The exactness of the binomial distribution's Critical Value (CV) is the main reason. One reason is that it could require a stoppage in shipments when the number of defects exceeds the CV. An additional reason is to have a monitoring system (diagnostics, cables, etc.) defect-free because one expects the HASA system to uncover only product-related issues and not deal with monitoring system issues as well. Defective products need to be the focal point of the process. More information on this can be found in chapter 4.

The original statistics anticipated more than five failures during the sampling period because the production processes were not within the statistical control that they needed to be. This "lack of control" allowed for the use of the binomial approximation. This proved to be beneficial because they learned about their production processes and the next obstacle that lay on their horizon: a six-sigma design. The six-sigma concept, introduced by Motorola in the late 1980s, incorporates both a quality goal and a methodology for achieving that goal. Specifically, the six-sigma quality goal can improve one's production processes so that the defect rate is no greater than 3.4 PPM (parts per million). The methodology to achieve this extremely high level of quality is predicate on two concurrent activities. The first activity is continually working to reduce process variation in manufacturing. The second activity is developing robust product and process designs that are little influenced by outside sources of variation. HALT and HASS are indispensable in both of these. Other tools that are also very helpful in achieving these goals are SPC (Statistical Process Control), Statistical Tolerance Analysis, and DOE (Design of Experiments). While a full treatment of the six-sigma concept and its associated tools is beyond the scope of this book, information on six-sigma can be found in many other textbooks. With six-sigma design and, as a result, vastly improved production processes, the details of chapter 4 were used some two years after the concepts set forth in this chapter. For most manufacturers who desire to perform stress auditing, this chapter will more than meet their needs.

Statistics—The System

The approximated statistical equation for the HASA process is,[3]

$$N = (Z_\alpha + Z_\beta)^2 \times p \times (1 - p) / D^2 \qquad \text{Eq 3.1}$$

where,
- N is the sample size to be stressed. The sample size N is calculated to provide an alarm point with confidence a so that corrective action can occur in a timely manner. One needs to assess not only the sample size N but to also consider the Days column in Table 3.2 which is discussed later.

- Z_α and Z_β are constants associated with the producer and consumer risks respectively. The producer's risk, α is the probability that a sample taken from acceptably good production (with failure rate p) will have sufficient defects that it appears to have come from unacceptable production (with failure rate $\geq p+D$). The consumer's risk, β, is the probability that a sample taken from an unacceptable production (with failure rate $p+D$) will have few defects so that it appears to have come from acceptable production (with failure rate $\leq p$). Low risks lead to higher values of Z_α and Z_β and correspondingly larger sample sizes. Table 3.1 shows Z-values for some common risk probabilities. Additional risk and Z values can be obtained from Z tables in many statistics books. Be sure to observe whether the Z value table from the book is α or 1-α.
- p is the baseline product historical failure rate percentage that can be the out-going failure rate.
- D is the shift in failure rate from p that is to be detected.

For a derivation of this equation, please see Appendix A.

The obvious effect that the failure rate shift has on the sample size can not be overlooked since it is typically a small number, possibly less than 0.1 (10%), squared, and in the denominator of the equation. On the other hand, the risk factors for α and β have an effect on the sample size but not quite as dramatic as D. Table 3.2 has some typical values for α, β, and D that result in a value for the sample size N. This table has an additional column labeled *# Defects* which is the number of defective units that potentially could be shipped before the shift D is actually detected by the stress system. Remember that only a very small sample of products are being stressed; hence, some defective products will ship before the first one(s) fails during stressing. The values in this column were calculated assuming a daily stress sample of 96 units and a daily shipment of 1000 units (not correlated with actual factory numbers). The equations for the *Days* column and the *# Defects* columns are:

$$\text{Days} = N \div 96 \qquad \text{Eq 3.2}$$

$$\text{\# Defects} = D \times \text{Days} \times \text{Ship Level} \qquad \text{Eq 3.3}$$

Table 3.1 Examples of α and β risks with their respective Z values.

Error	50%	25%	20%	10%	5%	2.5%	1%
Risk	0.500	0.250	0.200	0.100	0.050	0.025	0.010
Z Values	0.000	0.674	0.842	1.282	1.645	1.960	2.326

where,

Days are the number of days within which the failure rate shift, $D+p$, is to be detected. This interval can be overridden (Days + some value) if good engineering judgment is applied in determining the validity of the failure rate shift. This has been referred to as the sampling interval.

N is the sample size from Equation 3.1.

96 is the assumed number of samples stressed daily. This is dependent upon the number of units that the chamber can accommodate.

Defects are the increase in the number of defects that could be shipped in n Days due to a failure rate increase from p to $D+p$.

D is the shift in failure rate from equation 3.1.

Ship Level is the daily ship rate and is assumed to be 1000 for Table 3.2. Use a level appropriate for your production.

Equation 3.1 provides the sample size N given certain other factors. It does nothing to tell the user when to alert the corrective action team that a process shift has occurred. A control chart (see Figure 3.1) and good engineering judgment must be used. To further this discussion we will look at two scenarios: one involving the detection of multiple identical mode failures and the other involving multiple mode failures.

When the number of defects detected exceeds the alarm point, quick and correct action must be taken. The confidence with which this reaction is taken is directly attributed to $1-\alpha$. If the value of α were chosen to be 0.1 (10%), this would mean that when you make your decision to alert the corrective action team you could be wrong 10% of the time. This margin of error can be greatly reduced by quick and accurate failure analysis. With the failure analysis information and the α margin, you can be more accurate in your risk assessment. The value of α must be carefully chosen so that the correction action team is not erroneously activated. If this occurs, confidence in the HASA process may rapidly deteriorate.

Table 3.2 Sample size determinations for given conditions

α	β	Z_α	Z_β	p	$1-p$	D	N	Days	# Defects
.2	.2	.842	.842	.02	.98	.005	2223	24	120
.2	.2	.842	.842	.02	.98	.01	556	6	60
.2	.1	.842	1.282	.01	.99	.01	447	5	50
.2	.1	.842	1.282	.01	.99	.005	1787	19	95
.2	.1	.842	1.282	.02	.98	.005	3537	37	185
.1	.1	1.282	1.282	.01	.99	.01	651	7	70
.05	.05	1.645	1.645	.01	.99	.01	1072	11	110
.2	.1	.842	1.282	.01	.99	.02	112	2	40

Scenario 1

If failures of the same mode are detected within the time frame selected (the *Days* column in Table 3.2), then there is cause for concern. This indicates, even without detailed failure analysis, that a major issue has been uncovered. The corrective action team should be immediately notified. Provide them with all of the information that you have uncovered as well as any other pertinent, historical data you may have.

Scenario 2

If the failures are multiple modes, then a careful assessment must be made before activating the corrective action team. Failure analysis needs to be done so that good engineering judgment can be interwoven in order to provide an accurate decision. Here, the risk could be higher than when all of the failures are of the same mode. The following information should be sought:

- Are these new defect types?
- Are these defects singular in occurrence?
- Have these defects been previously "corrected"?

The information obtained from these questions can greatly enhance accuracy and determine your course of action since each requires a different reaction.

A Control Chart for the HASA Process

The control chart is an integral part of the HASA or HASS process and needs to be plotted as soon as samples are stressed. It should *not* be solely relied upon for the critical decisions. It can indicate trends, help in the corrective action assessment by providing timely information to production for new product introduction failure modes, and help in the overall decision-making process. The control chart only needs an upper control limit that is based on the sample size and the historical failure rate for the process. A lower control limit is of no value for this particular situation because when the process is below the upper control limit, it is in statistical control. Equation 3.4 was used to generate the values for the outgoing failure rate variable in Figure 3.1 (the dashed trace). Notice that the control limit changed immediately beginning at week 17. This was due to the fact that the failure rate was below the control limit for many weeks and a readjustment was deemed appropriate. This decrease in failure rate was a result maturing production which included more coverage for the board level test programs as well as better

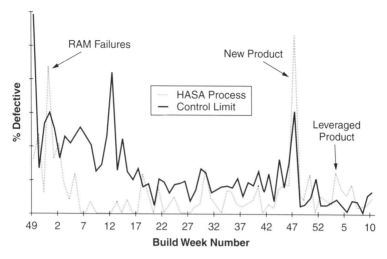

Figure 3.1 HASA control chart for products tested (excludes all test system and stress system problems).

assembly process documentation. Also notice that when a new product was introduced the failure rate increased beyond the control limit (week 47). Whenever this happened, the most likely culprit was the immature nature of the production test programs for that new product. Information was provided to production test and the problem(s) disappeared. Note, when the second new product was introduced three weeks later (it was a leveraged product in week 4 at the right side of the graph), its failure rate was significantly lower because of what was learned on the previously released product.

As mentioned in the previous paragraph, the plotted values for the outgoing failure rate were derived from a statistical equation. Here is that equation:

$$P_{success} = \frac{n!}{k!(n-k)!} P_{fail}^k (1 - P_{fail})^{n-k} \qquad \text{Eq 3.4}$$

where,

Equation 3.4 is the binomial distribution for the probability of k failures in
$\quad n$ trials and,
$P_{success}$ is the probability of success,
P_{fail} is the probability of failure,
n is the number of trials or tests,
k is the number of failures.

The Monitoring System Issues

Figure 3.2 is quite different from Figure 3.1. Figure 3.2 is used to monitor the failures that are monitoring system related. This graph depicts only failures

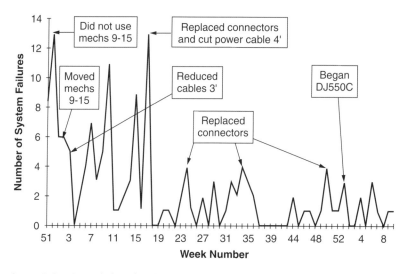

Figure 3.2 Control chart for all stress system and test system problems.

which were due to the test system such as connector contact, wire fatigue, diagnostics or self-test, the fixture, etc. Notice that every five to six weeks the failure rate increases. This is due to fatigue or wear-out of the interconnect hardware. Periodic replacement of this hardware is a must in order to have high confidence in the system. The word mech (as used in Figure 3.2) is the printer's print mechanism (the mechanical assembly which picks and moves the paper). The mechs were interconnected to the boards under test in the chamber through eleven-foot, Teflon jacketed cables. Additionally, the boards were connected to an industrial-grade PC which addressed each board and was, in turn, connected to a workstation which actually downloaded the diagnostics and stress profile to the boards under evaluation and the chamber controllers.

Of importance in this graph is the decrease in the mean time of the failures over time. In the beginning (the first one-third of the graph) the mean time was approximately three y-units. As time progressed (the middle third of the x-axis), the mean had dropped to approximately one y-unit system failure per week. During the end (the last third of the graph) the mean dropped even further—to about 0.5 units. This was a concentrated effort and the goal was to keep this low level to less than one failure per week.

Problems Uncovered through HASA

During the first two and one-half years in which the HASA process was in place, the following major problems were discovered and had corrective action implemented:

- During the early development phase of HASA, a problem with an Application Specific Integrated Circuit (ASIC) was uncovered. This problem only manifested itself during high temperature transitions above 55°C and during the high temperature dwelling at 75°C. Initially, it was thought to be related to a cable interconnection problem which had been seen previously and was thought to have been corrected. Failure analysis indicated, through the use of an acoustical microscope, that the three carriage motor drive leads in the ASIC had become delaminated. The fabrication facility for the ASIC was contacted, they replaced the molding machine, and the problem disappeared. This delamination occurred at the part's edge where the leads were attached and the concern about humidity penetration was raised.

- When the DeskJet 550C was subjected to HASA, it was noticed that the unit would behave abnormally anytime the temperature went below –17°C. This behavior was quite different from the previous printers that used the same technology and based upon this experience, it should have been able to operate to at least –50°C. After investigating, it was found that the input of a comparator had been left open. Its other input had an RC network. This product had been previously subjected to design review and STRIFE[L] and passed both. The RC network was added and the unit operated as its predecessors had done—down to –50°C.

- If the printer were turned on during the PC boot-up procedure, the printer would hang and require power cycling to clear the problem. This problem was traced to a firmware bug and corrected.

- Occasionally, three diodes would get misloaded in production and one day a good number were detected in HASA. The process engineering group found a software/machine interface problem that was corrected through software. This problem would become apparent to the user because the printer would *chatter* when no data was being sent.

- On the third generation of the ASIC mentioned in the first problem, a soldering problem was uncovered in HASA which passed the production bed-of-nails test. The placement of the part on the new board was rotated 90° and, because of the geometry and layout of the adjacent components, caused a soldering problem. This problem was also corrected.

- Early in the HASA development, a second problem was discovered that required a change in the HASA process. During an eight-hour period, 11 boards failed at 75°C. This problem was traced to the Random Access Memory (RAM) devices. They were removed and sent to failure analysis. While the parts were being evaluated, new RAMs were placed on the *suspect* boards, the boards were placed in the chamber, passed, and were subsequently shipped. When the results came back from failure analysis, they indicated that the RAMs were all good! Since the boards had been shipped, no further investigation of the boards was possible but eleven new boards were subjected to HASA, passed, replaced their RAMs with the evaluated RAMs, subjected to HASA, and passed. The HASA process was modified and included a quarantine of all assemblies when an issue of greater than three identical occurrences happened. The conclusion was that all of the original eleven boards had a soldering problem and not a RAM problem as was originally thought.

An Observation on Using Equation 3.1

As mentioned previously, the failure rate shift, D, which is to be detected has a large impact on the sample size N. This shift must be carefully chosen due to this impact. Notice that Z_α and Z_β also affect N. These are the risk factors derived from α and β and Table 3.1 has a range of Z_α and Z_β for risks (α or β) from 50 percent to 1 percent. The Z values for this entire risk range only goes from 0 to 2.326. This means that the values are going from a 50 percent risk to a 1 percent risk, and if we assume that $\alpha=\beta=0.01$, then the value of the term $(Z_\alpha+Z_\beta)^2$ goes from 0 to 21.6. With the higher Z_α and Z_β (lower risks) one can have a very high confidence and only slightly affect the sample size N. This fact would encourage one to design the statistical system with very low risks and exercise caution in the selection of the shift in failure rate.

Conclusion

Equation 3.1 provides a value for the sample size when the number of failures anticipated is greater than five. As the production processes improve, which takes time and a concentrated and focused effort, and the anticipated number of HASA failures becomes less than five, one must use the exact binomial equation. This situation is covered in the next chapter. Using the HASA system described in this chapter will help the user gain confidence and experience that will be essential in order to incorporate what is discussed in the next chapter.

HASA can be an extremely effective tool for the practitioner of accelerated techniques. It will, when properly tuned, detect shifts in the product's outgoing failure rate. It can be extremely accurate if the correct values for calculating the sample size are chosen. Risks are assumed on both sides, to the consumer and the producer, but both can be minimized with a slight increase in the sample size. The user must also realize that the production processes must be in statistical control for the full benefits of HASA to be realized. If not, then one must resort to using a stress screen or HASS, Highly Accelerated Stress Screen.

Additionally, HASA dramatically reduces production screening (over HASS) costs because of the reduced requirements for chambers, test equipment, personnel, utilities and floor space.

Four

Refinements on Highly Accelerated Stress Audit (HASA)

> To reach any significant goal, you must leave your comfort zone.
>
> **Hyrum W. Smith**

Introduction

As the overall production process(es) improve through the application of statistical methods, the number of defects which reach the field will decrease. Some of the assumptions made by this statement are:

1. The product was developed and ruggedized so that ample margins exist;[4]
2. A concentrated effort has been made to improve the production process(es); and
3. Highly Accelerated Stress Screen (HASS) has been in use and was later followed by Highly Accelerated Stress Audit (HASA) because the production processes are under such statistical control that five or fewer failures are anticipated over the HASA sampling period.

Complete product screens (HASS) are not bad, but because of their requirements they may create an adverse effect on profitability due to the following:

- *Product stressing systems.* HASS requires that all products be stressed and possibly more than one chamber be used. This is an expense that HASA could reduce because it uses an audit that may involve fewer chambers.
- *Manpower.* HASA requires less chambers, operators, and technicians.
- *Auxiliary test equipment.* Each chamber has its own test console or equipment to provide electrical stimuli to the product under evaluation and its associated cables. Sometimes, the cost of the test equipment may exceed the chamber cost.
- *Utilities.* Electricity and liquid nitrogen (LN_2) are required to heat and cool the product under evaluation. These expenses can be greatly reduced if an audit (HASA) is used. The cost of the additional LN_2 piping can also be reduced through the implementation of an audit.
- *Employee training.* Techniques are improving and technology is advancing. As a result, each employee involved in operating the chamber should receive training. If there are more chambers, more staff training is required.

If these costs are significant to you, this chapter will help you reduce costs through a process called HASA. A form of HASA was presented in chapter 3. It performed extremely well at Hewlett-Packard's (HPs) Vancouver Division on the DeskJet family of printers during the first two years of product stressing. The rationale for changing the original HASA process to the one presented herein was due to the fact that the production processes improved to the point where fewer than five failures were being anticipated and detected during the sampling period. At this high level of quality, it became necessary to use a refined statistical system that provided an exact binomial expansion rather than an approximation. One can begin applying the HASA techniques by using those described in chapter 3 or by adopting those detailed in this chapter—either can meet your needs. Alternatively, an additional, more refined process that has been in use since January 2000 is covered in the last half of this chapter.

Background and Assumptions

The statistics used in this chapter have been simplified so that the reader will be able to "plug in the numbers and get results." For those who have limited familiarity with statistics this, of course, is ideal. Every stress screening process needs to eventually perform audits rather than merely screening every product. This can greatly reduce the overall cost for product stressing and increase profitability. There may be contractual constraints against auditing but for this chapter the assumption is made that they do not exist.

We will assume that the production processes and supplier's processes are within statistical control. In other words, a working relationship has been established between the manufacturer and the supplier and, because of the high level of part quality, incoming inspection at the manufacturer is usually not required. We will also assume that continuous process improvement projects are ongoing. The processes are focused on providing defect-free products when the process parameters are optimally set. With these mechanisms in place, a stress process may be changed from a screen (when 100% of the products are stressed) to an audit (a statistical sample is stressed). This audit does not require the stressing of 100% of products when failures occur. Today we use statistical methods such as: Total Quality Control (TQC), Total Quality Improvement (TQI), Continuous Quality Improvement (CQI), Taguchi, six-sigma designs, and others. These tools and methods are much more effective than those used during the early days of MIL-STD 105 in measuring and improving products and the processes that produced them.

Application of the Statistics

The statistics for this chapter are shown in Equations 4.1 and 4.2. These equations are used to calculate the sample size, n (N in chapter 3), and the critical value, c.

$$\sum_{j=0}^{c} \frac{n!}{j!(n-j)!} p^{j_{acc}}(1 - p_{acc})^{n-j} \geq 1 - \alpha \qquad \text{Eq 4.1}$$

$$\sum_{j=0}^{c} \frac{n!}{j!(n-j)!} p^{j_{rej}}(1 - p_{rej})^{n-j} \leq \beta \qquad \text{Eq 4.2}$$

where,
- n is the sample size to be stressed.
- j is the number of failures of n items.
- c is the critical value (CV). This is the limit of the number of failures during stressing that are allowed. A failure level of CV+1 may require the activation of the corrective action team.

- α is the producer's risk. An improper choice in the selection of this parameter may mean that you may reject a "good" lot. In other words, if $\alpha=.05$, you have a 5% chance of rejecting a good lot of p_{acc} quality.
- β is the consumer's risk. Choosing a $\beta=.05$ would mean that 5% of the time you will accept a lot with p_{rej} defective.
- p is the baseline failure rate of the product, where,
  - p_{acc} is the quality of the lot fraction defective considered acceptable,
  - p_{rej} is the quality level at which we reject a lot ($p_{acc} + D$),
- D is the shift in the failure rate which we wish to detect or $p_{acc} + D = p_{rej}$.

For the sake of the user and simplicity, Equations 4.1 and 4.2 have been solved to provide the values shown in Table 4.1. This table shows some typical values for α, β, and p, the corresponding values for the sample size n, and the results of calculating the sample size by using the exact binomial distribution Equations 4.1 and 4.2. An example on how to use Table 4.1 would be:

- The consumer's risk α, is 0.05 (5%);
- The producer's risk β, is 0.10 (10%);

Table 4.1 Samples sizes for given α, β, p, D and CV.

α	β	p	Sample Size for D					Critical Value for D				
			0.01	0.02	0.03	0.04	0.05	0.01	0.02	0.03	0.04	0.05
0.05	0.10	0.005	783	266	151	117	70	7	3	2	2	1
		0.010	1235	390	198	132	110	18	7	4	3	3
		0.020	2079	616	306	194	131	52	18	10	7	5
		0.030	2872	807	410	252	175	101	32	18	12	9
		0.040	3667	1001	496	292	208	166	50	27	17	13
		0.050	4445	1196	572	348	233	246	72	37	24	17
0.10	0.20	0.005	447	171	85	66	54	4	2	1	1	1
		0.010	681	223	106	85	50	10	4	2	2	1
		0.020	1075	312	157	111	78	27	9	5	4	3
		0.030	1525	427	207	129	98	54	17	9	6	5
		0.040	1933	537	257	155	113	88	27	14	9	7
		0.050	2337	628	307	187	124	130	38	20	13	9
0.20	0.20	0.005	285	119	85	66	54	2	1	1	1	1
		0.010	453	142	74	59	50	6	2	1	1	1
		0.020	713	226	110	71	60	17	6	3	2	2
		0.030	967	272	150	95	68	33	10	6	4	3
		0.040	1219	337	161	98	74	54	16	8	5	4
		0.050	1488	413	197	125	78	81	24	12	8	5

- The base line failure rate or acceptable quality level, p_{acc} is 0.010 (1%); and

- The shift in the outgoing failure rate, D or $p_{acc} + D = p_{rej}$, to be detected is 0.030 (3%).

For these values, read across using the second row under α (0.05) and β (0.10) and p (0.010). The sample size, n, is 198. If we had used the sample size equation in the previous chapter (Equation 3.1) to derive the sample size n, it would have yielded 95 samples. Keep in mind that this chapter utilizes statistics that are more precise than those used in chapter 3.

Table 4.1 has a set of five columns labeled as "Critical Value for D" which is the number of failures that can be *accepted* during the production stressing. The CV becomes the acceptance criteria for a given sample size, n. The CV columns in this application replace the need for the HASA control chart that was used in the previous chapter although it can be used. Each of the five CV columns corresponds to each sample size column. If the number of failures exceeds the CV, as in the case of the second row, then the statistics indicate that the failure response team *should* be activated. The following question needs to be addressed before activating the failure response team: Are all of the failures different or do they have commonality? This is a crucial question to answer because the stressing process will uncover failures but the failures may not have any modes in common. If they are not related, you may want to consider not activating the team. On the other hand, you may want to activate the team if there is more than one common failure or if the failure modes were once driven to root cause and corrected. One may also want to investigate if the CV is not exceeded but all of the failures are the same. In all cases, one should carefully review all failures.

After examining the same five failures (CV+1) from Table 4.1 graphically in Figure 4.1, we can see that these failures did come from rejectable production. Any number of failures in excess of the CV need to be analyzed for possible corrective action. The CV+1 value shown in this example is the white bar at the 5 failure level on the x-axis and all failures to the right of this value. On the other hand, if the number of failures is less than or equal to the CV, continue the stressing process. This situation is indicated in Figure 4.1 by zero through four failures.

If the sampling plan is less stringent than the one that was just used (higher risks for α and β), then the graphics change. Figure 4.2 shows such a situation. In this case, the resulting sample size is 106 and the CV is 2. Once again, the failure team is activated only when the number of failures is CV+1 or 3. Notice that the CV is lower because the sample size has decreased due to the higher α and β risks.

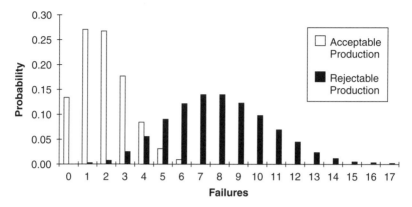

Figure 4.1 Probability of failures for acceptable and rejectable production for $N=198$, $\alpha=0.05$, $\beta=0.10$, $p=0.01$, and $D=0.03$.

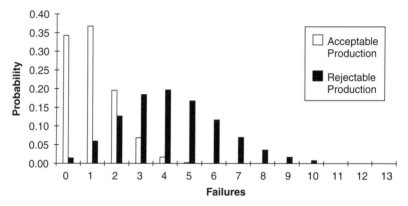

Figure 4.2 Probability of failures for acceptable and rejectable production for $N=106$, $\alpha=0.10$, $\beta=0.20$, $p=0.01$, and $D=0.03$.

A Graphical Tool for Detecting Defect Level Changes

This section may be used as an analysis tool. Cumulative sums can be used as a tool to detect defect level changes. The sample sizes and CVs given previously are useful for detecting specified shifts in the defect level. However, these sample sizes can be quite large and the tests do not provide signals when there are smaller shifts in the defect level. One technique for making shifts in the defect level visible is to plot the cumulative sum of the defects against the total units tested. The cumulative sum is typically adjusted to account for an expected rate of defects (using historical records, targets, or

Table 4.2 Historical HASA data used to generate Figures 4.3 through 4.5.

Number Tested	Defects (Fails)	Cumulative Tested	Cumulative Defects	Adjusted Defects
2	0	2	0	-0.004
32	1	34	1	0.932
106	5	140	6	5.720
90	1	230	7	6.540
80	7	310	14	13.380
96	3	406	17	16.188
145	6	551	23	21.898
123	2	674	25	23.652
138	1	812	26	24.376
124	3	936	29	27.128
112	0	1048	29	26.904
123	0	1171	29	26.658
140	1	1311	30	27.378
223	0	1534	30	26.932
117	0	1651	30	26.698
125	0	1776	30	26.448
69	0	1845	30	26.310
148	1	1993	31	27.014
183	0	2176	31	26.648
114	1	2290	32	27.420
135	0	2425	32	27.150

the specified acceptance quality level). An example using data from actual screens is shown in Table 4.2.

These data represent only a segment of the historical record for this product. The first two columns list the number of units tested and the observed failures for this product. The next two columns show the cumulative sum of both the units tested and the units failed as calculated from the first two columns. The last column is calculated from the third and fourth columns as cumulative defects or *Cumulative Defects – 0.002 × Cumulative Units Tested*. The reason for choosing the multiplier 0.002 will become apparent as we look at a plot of these data.

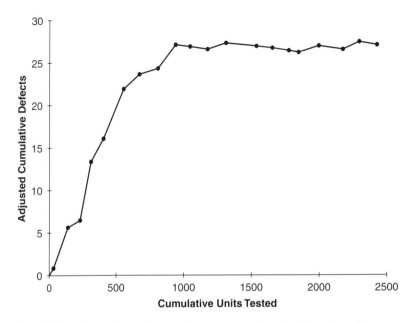

Figure 4.3 Adjusted cumulative defects versus units tested showing solder process defect correction.

The sample plot in Figure 4.3 shows that the adjusted cumulative defects have remained constant at about 27 defects after 1000 units had been tested. When the plot of the adjusted cumulative defects is horizontal (flat), the defect rate for that period is equal to the adjustment factor (here, 0.002 or 0.2%). When the plot has a positive or negative slope, this slope represents the difference between the defect rate and the adjustment factor. For this plot, the curve is nearly linear for the first 1000 units tested. The slope in this period is approximately 27/1000 (0.027 or 2.7%). These data clearly show that the defect rate changed after approximately 1000 units had been tested. Indeed, the high failure rate of the early test units was traced to a soldering problem on a specific component. The soldering process was improved and the defect rate was correspondingly dropped. Notice the flatness of the plot between 1000 and 2500 units tested. This indicates the effectiveness of the corrective action.

Additional units were tested (beyond Table 4.2) and Figure 4.4 shows the adjusted cumulative defects for the first 8000 test units (the adjustment factor, $p=0.002$). Notice the jump in the curve at about 2500 units and a positive slope after about 4800 units. When the adjusted cumulative defects curve jumps (as it does at 2500 units) there is an unusually high number of defects for those units tested.

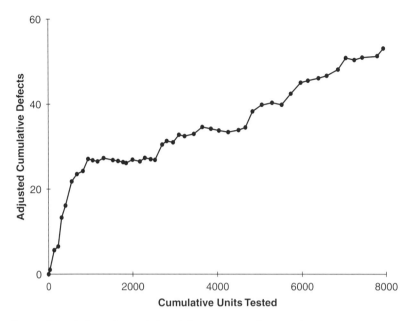

Figure 4.4 Adjusted cumulative defects versus units tested.

The curve is nearly flat (perhaps with a slightly upward slope) between 2500 and 4800 units, showing that the failure rate for these units is only slightly larger than the 0.002 adjustment factor. For one sample at 2500 units, however, the defect rate seems uncharacteristically large and might require further investigation.

The failure rate for the last 3200 units seems larger than the 0.002 adjustment factor. By plotting the same data with various adjustment factors, one can find the defect rate that approximately matches the data. For these data, an adjustment factor of 0.0065 seems to be appropriate. Note that the negative slope between 1000 and 2500 units tested in Figure 4.5 indicates that the adjustment factor of 0.0065 is too large for this period. In addition, the adjustment factor seems too large for the units between 3800 and 4800. There also appears to be a sample at 4800 units with a large number of defects.

Plotting adjusted cumulative sums is an effective way to show changes in failure rates without getting too caught up in sample sizes and CVs. Defect rates can be readily estimated by trying various adjustment factors until a horizontal curve in the region of interest is obtained. Jumps in the curve show one-time increases in the defect rates whereas linear slopes show a constant failure rate that is different than the adjustment factor by an amount equal to the slope. See Figure 4.5.

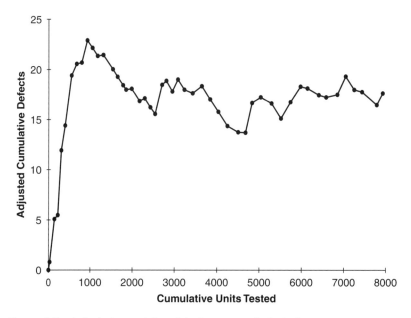

Figure 4.5 Adjusted cumulative defects versus units tested.

Conclusion

It has been shown that setting up a statistical system for calculating the stress sample size can be easily done. The user will need to make careful determinations on two risk factors: α and β. It is important to realize the implications if these factors are incorrectly chosen. If the overall failure rate, p, is not precisely known, one may use the failure rate of a previous product (or make an educated guess) that is similar to the new one. Appropriate adjustments to p because of the enhancements to the new product and the sensitivity or shift variable D will need to be considered. Table 4.1 is a good starting point for CVs without getting involved with the graphical tools that precede these conclusions. For the user that feels comfortable with the beginning of this chapter and would like to view the same statistical system from a graphical perspective, then the section entitled "Application of the statistics" immediately following the opening section of this chapter will be of value.

Introduction to an Improved HASA Process

HASA and HASS are methods used to detect the presence of manufacturing defects and prevent products with such defects from being shipped to the field. The type of defects addressed in HASS and HASA are those that occur in the early life of a product. By eliminating the units from the field population that are prone to early life failure, the infant mortality portion of a product's reliability curve can be significantly reduced, if not eliminated.

As stated previously, HASS is a screening procedure. This means that 100% of the products are tested. HASA is an audit or sampling procedure. This means that less than 100% of the products are tested. The purpose of this section is to describe another sampling process for HASA that can be used in place of 100% inspection and to describe the performance of and risks associated with such an approach.

While the primary purpose of HASS is the *detection* of early life failures and the "screening out" of such products from the field population, an equally important feature of HASS is that it provides necessary information for process monitoring and improvement. The emphasis for HASA however, is slightly different than in HASS. The primary purpose of HASA is process monitoring and the *prevention* of manufacturing defects. The sampling plan for HASA is designed to quickly signal when a degradation in process quality has occurred.

Once such a signal is given, the process corrective action team can be activated to identify the cause of the degradation and return the process to its normal operating quality levels. Unfortunately, HASA does not have the ability to completely prevent defective products from being shipped. Since some of the boards under HASA do not undergo testing, it is inevitable that some defective products will be shipped to the field. This is the a fundamental difference between HASS and HASA.

HASA Process Flow

The HASA and HASS process flow could begin when products arrive in the HASS test area. At this point, if the capacity to allow 100% inspection exists then all products are delivered to the HASS area and subsequently tested. If there is not a sufficient capacity to allow 100% inspection then the products are sampled and only a fraction of them are tested. Those units not selected for the sample bypass HASA testing and proceed through the remainder of the process. When a sufficient number of products have been tested (as stipulated by the HASA sampling plan) then the statistical sampling plan is applied to the data and the appropriate conclusion is made.

Under this sampling plan, one of three possible conclusions will be made when the statistical sampling plan is applied to the data. If the data indicates that the quality level of the process is acceptable then the conclusion is that the process is OK and no action is required. As a result, the HASA process continues. If the data indicates that the process quality level has degraded to a level that is not acceptable, then one of two remaining conclusions will be made: the process is either at *Quality Level I* or *Quality Level II*. By definition, both Quality Level I and Quality Level II represent a degradation in process quality below what is acceptable. However, Quality Level I is not bad enough to justify a stop production or stop shipment order. Quality Level II represents a more severe degradation in the process quality level and would justify stop shipment and stop production orders.

If the data indicates that the process is operating at Quality Level I then the process corrective action team is activated to identify the cause of the quality degradation. Once identified, the issue is entered into a defect tracking system (DDT, see chapter 1) and a corrective action team is activated so that the problem can be fixed as quickly as possible. No change in the shipping or production schedules will occur.

If the data indicates that the process is operating at Quality Level II then, in addition to activating the corrective action team, shipping and production are stopped. The number of defective products being produced is too large to allow shipping or production to continue. The HASA process flow is summarized in the flow chart shown in Figure 4.11 on page 85.

Typical Lot Acceptance Sampling Plan

A typical "Lot Acceptance Sampling Plan" consists of a sample size (n) and a decision limit (c). When applying the sampling plan to a set of data, a sample of size n units is taken from the lot (or process) and the number of defective products counted. If c or fewer defective products are found in the sample of n units then the lot (or process) is accepted. If more than c defective products are found in the sample then the lot is rejected and appropriate action is taken.

Tools used to describe the "performance" of an acceptance sampling plan (and the risks involved) are the Operating Characteristic (OC) Curve and the Average Run Length (ARL) Curve. The OC Curve shows the probability of accepting a lot for various levels of lot fraction defective. Clearly, if a lot is of a high quality level then we want to have a high probability of accepting the lot and a low probability of rejecting the lot. And if the lot is of a low quality level then we want a low probability of accepting the lot and a high probability of rejecting the lot. The OC Curve shows these probabilities.

The ARL Curve shows the average number of lots (for various lot quality levels) that will be submitted to the plan before a lot is rejected. If the lot is of a high quality level then we want a large average run length. In other words, we want to run for a long time before a lot is falsely rejected. On the other hand, if a lot is of a low quality level then we want a small average run length. In other words, we want the sampling plan to quickly reject a lot.

As an example, consider a sampling plan with $n=100$ and $c=3$. Shown in Figures 4.6 and 4.7 are the OC Curve and the ARL Curve that describe this plan. For this specific plan, the OC Curve shows that the probability of accepting an individual lot with .04 fraction defective is approximately .42. The ARL Curve shows that if repeated lots having .04 fraction defective are submitted to the sampling plan, it will take 1.8 lots on average before a decision to reject is given.

HASA Acceptance Sampling Plan

The HASA sampling plan is a modification of a typical lot acceptance sampling plan in that it has two decision limits: c_1 and c_2. The reason for using two decision criteria is to allow for more flexibility in the type of action that will be taken if the decision criteria are exceeded. When applying the HASA sampling plan to a *production lot* (the process stream), a sample of size n units is taken from the process stream over the period of time specified by the plan. The term *lot* is used loosely here and is used to represent that subset of the production stream from which the sample size n is collected. In the plan described below, a *lot* corresponds to all products produced in a single day. Once the n units are selected from the lot, they are tested in HASA and the number of defective units are counted. If c_1 or fewer defective units are found in the sample then the process is accepted. If exactly c_2 defective units are found in the sample then the process

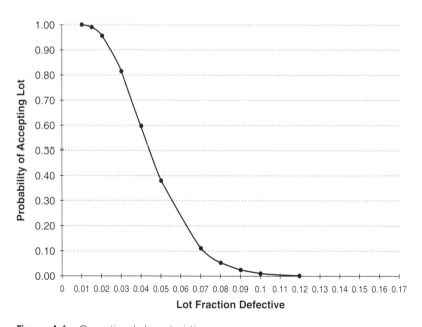

Figure 4.6 Operational characteristic curve.

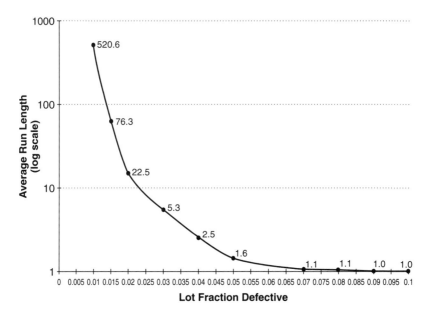

Figure 4.7 Average run length curve.

is deemed to be at Quality Level I and appropriate action will be taken. If more than c_2 defective units are found in the sample, then the process is deemed to be at Quality Level II, and appropriate action will being taken.

Given a daily HASS capacity of, say, 128 units per day, the following sampling plan has been established for these units: $n=128$, $c_1=5$, and $c_2=6$. This plan was set up so that one day's production constitutes a "lot" and a decision regarding the process will be made once per day. In considering quality levels, a process operating at a level of 1.5% defective or less is considered acceptable and a process operating at a defect level higher than this is not acceptable. The OC Curve and the ARL Curve for this plan are shown in Figures 4.8 and 4.9, along with the tabular form of the data in Figure 4.10. It can be seen that for this plan, if the process is operating at 0.015 fraction defective, the probability of accepting the process is 0.987 and the average run length for a process at this level is 76.3 lots.

This plan is an example of the approach that could be used. However, the specific characteristics of the sample plan (sample size and decision criteria) could change, depending on product type and HASS capacity. Furthermore, a separate sampling plan could be established for high volume and low volume products.

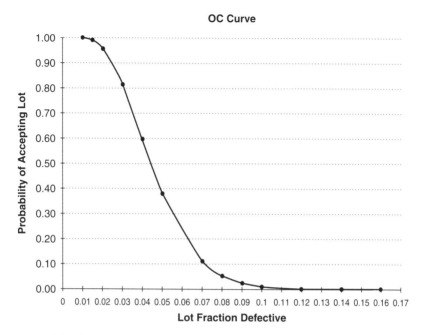

Figure 4.8 Operating characteristic curve.

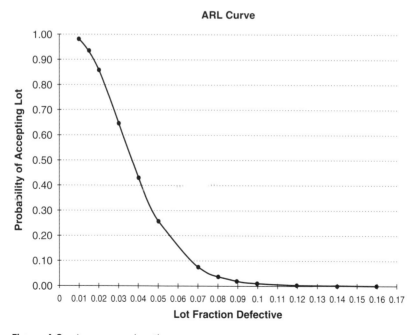

Figure 4.9 Average run length curve.

84 Chapter Four

HASA Sampling Plan for High Volume Products

Sample Size: n = 128
Decision Criteria: 5 or fewer defects = Accept Process
6 defects = Reject Process; Quality at Level I
7 or more defects = Reject Process; Quality at Level II

Actual Process Fraction Defective	Sample Size n	Defects Found =	Probability of..... Accept Process 5	Reject Process at Level I 6	Reject Process at Level II 7	Average number of Lots inspected before rejecting the process	Average number of Lots inspected before a Level II signal occurs
0.01	128		0.998	0.002	0.000	520.6	3034.9
0.015	128		0.987	0.010	0.003	76.3	299.5
0.02	128		0.956	0.030	0.015	22.5	67.1
0.03	128		0.812	0.096	0.092	5.3	10.9
0.04	128		0.595	0.153	0.253	2.5	4.0
0.05	128		0.378	0.162	0.459	1.6	2.2
0.07	128		0.109	0.091	0.799	1.1	1.3
0.08	128		0.052	0.054	0.894	1.1	1.1
0.09	128		0.023	0.029	0.948	1.0	1.1
0.10	128		0.009	0.014	0.977	1.0	1.0
0.12	128		0.001	0.003	0.996	1.0	1.0
0.14	128		0.000	0.000	0.999	1.0	1.0
0.16	128		0.000	0.000	1.000	1.0	1.0

Figure 4.10 HASA sampling plan for high volume products

Refinements on Highly Accelerated Stress Audit (HASA)

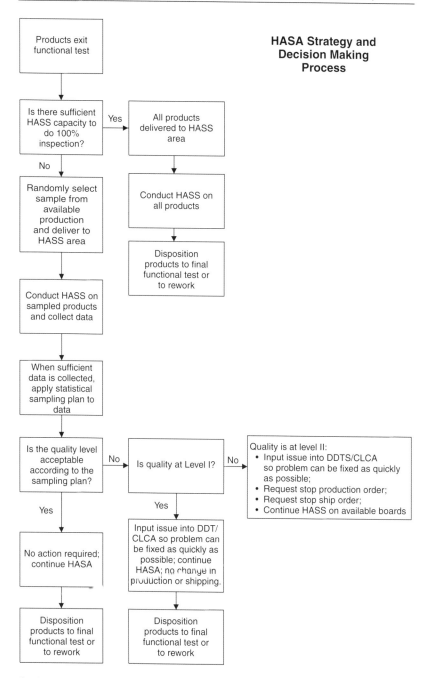

Quality Level I—represents a degradation in the quality level below what is acceptable. However, the quality level is not low enough to justify a stop production or stop shipment order.

Quality Level II—represents a degradation in the quality level below what is acceptable and below quality level I. Quality level II is low enough to justify stop shipment and stop production orders.

Figure 4.11 HASA decision making

FIVE

The Equipment Required to Perform Efficient Accelerated Reliability Testing

> A new scientific truth does not triumph by convincing its opponents and making them see the light, but rather because its opponents eventually die and a new generation grows up that is familiar with it.
>
> **Max Planck**

Overview

In order to attain the high reliability that has been mentioned in the previous chapters, the proper selection of equipment is necessary. Improved reliability *can* be achieved using traditional methods but not to the high level that is attainable by using the equipment discussed throughout this book. Recall that in chapter 3, Hewlett-Packard's (HPs) DeskJet facility was not able to mimic their customer's experience using traditional equipment. As stated in the glossary of this book, the definition of traditional equipment includes electrodynamic vibration, slow rate of change temperature chambers (compressor-based systems), and chambers not capable of performing combined stresses (vibration and temperature). Therefore, the equipment outlined in this chapter may seem non-essential to the novice but it has been used throughout the 1980s and 1990s and is gaining even more acceptance in the new century for the HALT, HASS, and HASA processes.

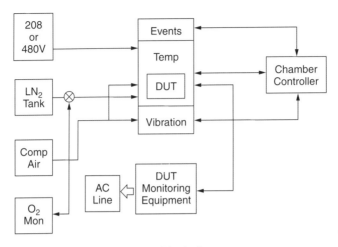

Figure 5.1 Accelerated stress system block diagram.

Figure 5.1 is a block diagram of a typical accelerated stress system. The chamber is a combined temperature and vibration system. The details of these systems will be presented later in this chapter. The oxygen depletion O_2 monitor should be capable of stopping the flow of the liquid nitrogen (LN_2) when a nitrogen alarm occurs by activating an automatic shutoff valve on the LN_2 tank. This valve will only open and allow LN_2 to flow if the alarm condition no longer exists. Consider multiple monitors if the chamber is to be located in an enclosed space rather than in an open area like a production floor. Figures 5.2 and 5.3 are photos of chambers that are typically used for HALT, HASS, and HASA. Other models are available from both manufacturers.

Temperature

Temperature is an important stress which causes latent flaws to become patent failures for many products. Some product's flaws are time (t) sensitive (dwell or $t_1 - t_2$); others are rate or a change of temperature (T) versus time sensitive ($\Delta T/\Delta t$); and others are temperature range sensitive (T_1 to T_2). It is, therefore, imperative that when designing the stress regimen that all of these stresses be considered. Compromises may have to be made in order to have a cost-effective, robust screen. On the other hand, long temperature dwells do little to cause failures to reveal themselves and these dwells should only be as long as it takes the product to perform its diagnostics and for product thermal stabilization.

Figure 5.2 OVS-2.5eHP *(Courtesy of QualMark Corp.)*

Figure 5.3 QRS-600V *(Courtesy of Screening Systems Inc.)*.

Turbulence

Turbulence (or uneven airflow) is an essential component of an accelerated stress system. With turbulence, the component's thermal transfer characteristics are greatly improved. Laminar airflow (or even airflow) on the other hand, is not recommended for accelerated stressing. It does not break down the air pockets which can be created around components and causes hot or cool spots on the item being evaluated. With turbulent airflow, this condition does not exist.

Heating

The temperature changes in these chambers are created by two different means: alternating current (AC) and liquid nitrogen, LN_2. Heat is created by the AC passing through many banks of multi-phase heaters. There is usually at least one fan that creates turbulent air, which in turn passes around the heaters, forcing the heated air into the chamber cavity and onto the product. The air passes through the product and returns to the heater and fans and is continuously recycled. It is not uncommon to have turbulent airflow in excess of 2700 cubic feet per minute (CFM) in these systems. Compressor-based thermal chambers have an airflow that is much lower than these levels. As a result, their lower product temperature change rates.

Cooling

The cooling system uses the same fan(s) as the heating system but instead of using a refrigeration compressor for cooling, it uses LN_2. LN_2 is dispensed through a series of nozzles internal to the chamber which are in the same air path as the heaters. As the LN_2 is dispersed through these nozzles, it immediately turns to a cold gas (-193°C). This gas is then circulated in the same fashion as the heated air. As this gas expands it creates a positive pressure difference inside the chamber it exits through the exhaust port(s) in the chamber. This gas should be vented outside the building for safety reasons. It is also recommended that at least one oxygen depletion monitor be mounted on a wall near the chamber so that personnel can be alerted in the event of a nitrogen leak and the supply of LN_2 can be automatically interrupted. These monitors should be located on the wall at about sitting height or approximately 30 inches from the floor. Personnel evacuation from the work area should be quick since LN_2 is tasteless, odorless, cannot be seen, and can cause death by asphyxiation through prolonged exposure.

A Comparison of LN_2 Systems and Compressor Systems for HALT and HASS

In spite of some of the shortcomings of the LN_2 system stated in the previous paragraph, its benefits can be enormous. Due to its simple design, the LN_2 system has a much better field performance record than its compressor counterpart. When compressor systems are pushed to their limits, they tend to fail at a much higher rate than when performing slow temperature change rates with long temperature dwells. This is exactly opposite to what is required for an effective HALT, HASS, or HASA. Since time compression is one of the goals of HASS or HASA, many more products can be stressed without causing the LN_2 system to even work hard. As a result, it is appro-

Table 5.1 Comparisons of compressor system and LN$_2$ type system

Parameter	LN$_2$ System w/Vib	Mechanical System w/o Vib
Expenses		
Purchase	1x	2.5x
Installation	About the same	About the same
Operational	1x	6x
Maintenance	1x	200x
Reliability	Excellent	Very poor
Size	1x	2x
Weight	1x	2x
Noise	60dBA	70 to 85dBA
Thermal		
Range	-100°C to 200°C	-40°C to 177°C w/cascading
Ramp Rate	(60°C per minute)	(30°C per minute)
Safety	Vent to outside Displaces oxygen	Possible, phosgene gas

priate to include a comparison of the two systems using the LN$_2$ system as the basis for comparison. See Table 5.1.

The shortcomings of the compressor system for HALT and HASS applications are evident in Table 5.1. The operating factor of 6 times is a nominal comparison since utility costs are extremely variable from location to location as well as one's ability to negotiate the price of LN$_2$. The phosgene gas listed under mechanical system is a by-product of heating the compressor pipes without first evacuating them of freon. Trained mechanics are aware of this danger and therefore, take precautions to eliminate this hazard by evacuating freon from the system first. This hazard is not unlike a nitrogen leak or the improper exhaustion of nitrogen with the LN$_2$ system.

An additional feature of the LN$_2$ chamber is its ability to have a higher product throughput. This throughput for the LN$_2$ system could be about 24 times more efficient than the mechanical system. See Figure 5.4.

The 24 times for throughput was derived as follows. Using the thermal limits from the graphs in Figure 5.4 and 10 minutes dwells, the slower chamber would have a 160 minute cycle time versus about a 24 minute cycle time for the LN$_2$ chamber. The ratio of these two cycle times is about 7. Running each of the two profiles through an acceleration equation yields 5 and 17 respectively. The value of 24 is derived finally by, $7 \times 17/160\text{min} = 1 \times 5/160\text{min}$.

The compressor system costs less to operate if it is operated in its normal (slow) mode of operation performing repeated temperature cycles with

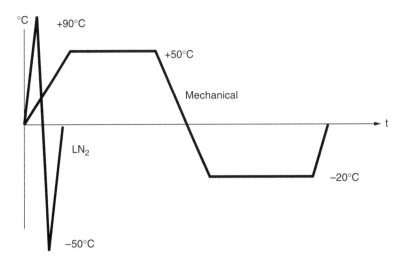

Figure 5.4 Using each type system as they were designed (optimized).

lengthy temperature dwells. The LN_2 system far exceeds the compressor system temperature performance and is able to uncover the same defect type in far less time. In other words, what may take the compressor system roughly 24 thermal cycles to uncover, could conceivably be done in one cycle using an LN_2 system because of its rapid temperature transition rate and wider temperature range. This is a dramatic difference and needs to be considered when deciding on thermal systems. LN_2 systems have been optimized for rapid temperature ramps and short dwells and when used as designed are extremely cost effective. This time compression is due to the exponential fatigue damage shown in Equation 2.1 (see chapter 2).

Reference was made to reliability and Table 5.1 indicates that there is a 200 times difference. This value is conservative but realistic when considering the replacement of compressor parts and systems when they are pushed to their limits. Consideration of down time is also essential because parts typically need to be ordered which can be time-consuming. There are many components in a compressor-based cooling system but only two in a typical LN_2 system—the proportional valve and a redundant safety valve. (As a side note, two experienced chamber technicians who had worked with both system types supplied the actual difference of 200 times.)

Vibration

The vibration systems used in HALT, HASS, and HASA are not the traditional electrodynamic or mechanical style systems. The recommended vibration system utilizes a six degrees of freedom style pneumatic system. These tables vibrate in the three orthogonal axes and three rotational vectors using each of the three orthogonal axes simultaneously. Independent control over each of these vectors is not required. The axis correlation on these styles of tables will indicate that there is an axis imbalance. However, some tables have much less balance than others do. A table that has the same energy in all three orthogonal axes would indicate that there should be little concern for the user regarding whether the product under test should be fixtured in the x, y, or z axis. The second mode can be seen by using a strobe in the unsynced mode while looking at the product being vibrated. This is created by the angular vectors that exist in many products' end-use environments.

The pneumatic vibration system is capable of producing vibration in six degrees of freedom that are random in nature. In other words, the stimuli are non-coherent and non-stationary, thus producing random vibration.[8] Some vibration systems have a spectral smearing feature that helps to excite product modes. This seems better than a picket-fence type spectrum which may not excite certain modes because no energy is present between the pickets or worse, over excites resonances where a spike of vibration is present. All of the tables that use pneumatic vibrators also have higher vibration levels at the table edge than in the center because of a twisting motion in the shape of a saddle (also known as the saddle effect). With this in mind, mapping the vibration response over the entire table surface may not be a meaningful effort. A word of caution—if multiple products are to be stressed simultaneously (HASS or HASA) then each product position *must* have a proof of screen performed. In other words, if there will be 10 positions for products to be mounted on the table, then each of the 10 must have a proof of screen done. See chapter 2 for more details regarding the HASS process.

The pneumatic vibrators are mounted under the table so that the product under test will have energy coupled to it through the fixture from all six axes simultaneously. Figure 5.5 is an example of a Power Spectral Density (PSD) from one of these types of tables. The lowest of the four plots does not have an accelerometer connected to it and is open (does not display meaningful data). The three top plots display the vibration response on three products located in a chamber with an accelerometer on each in the same axis. Each plot has frequency on the x-axis and G^2/Hz on the y-axis with the vibration response indicated within each plot in Grms. Some tables have energy from 2Hz to well above 5KHz. The low frequency energy is important for exciting components with a high-mass while the high frequency may be useful for exciting the modes of low mass Surface Mount Technology (SMT) components. For those not wishing to have the high frequency energy transmitted to their product, the fixture can be designed to decouple this energy. It is important to realize that the product and its fixture will attenuate vibration energy regardless of the energy spectrum that is present at the table. It is also

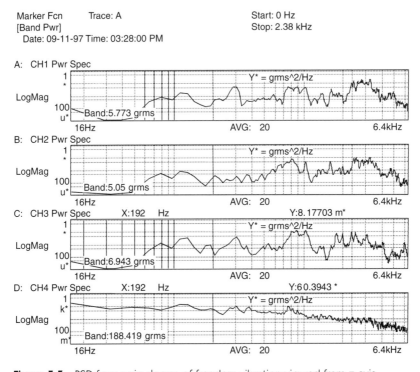

Figure 5.5 PSD from a six-degree of freedom vibration viewed from z-axis.

important to consider the limited bandwidth (up to 2 KHz) on some vibration systems. This may become a limiting factor as your products become more and more robust and there may be real failure modes that can only be excited with the wider bandwidth and higher energy levels. The broader vibration ranges help to excite high-mass components, small SMT device internal bonds, and external solder joints. For those who are accustomed to traditional electrodynamic shaker PSDs, the six degrees of freedom types of PSDs may seem uneven and even of little use for product evaluations. These tables have not only been used in the normal HALT environments but also fixtures have been designed that can cause these types of tables to provide a NAVMAT-type PSD. This is so that the user will not have to rotate the product between shakes on each axis and therefore realize a 66% time savings.[7]

For those companies that have a compressor-based system that can accommodate a vibration table (AGREE style chamber), it is highly recommended that a six degree of freedom vibration table be added to their compressor-based system. The thermal system is not the ideal solution but the results obtained with the improved table design are far superior than using a traditional vibration table.

Having said that the ideal vibration solution is the six degrees of freedom style table, it is important to note that it is not a panacea (it may not uncover all of the possible vibration induced defects). There may still be a need to use the traditional vibration systems when large displacements are required to induce failures on high-mass components. I would recommend that one perform HALT using the six degree of freedom table first, ruggedize the product, and then use the traditional vibration system.

Control Systems

The control system comes in many styles and packages. The key items for a control system, assuming that it performs its basic function of thermal and vibration control, are:

- *User friendliness.* Are there many levels of menus that don't allow the user to escape?

- *Speed.* Does the stress system or graphical computations visibly slow down the processor?

- *User configurable.* Can the user configure colors and the user interface? Can the software be upgraded to include vibration analysis? Can the system plot both data for HALT and repeated profiles such as HASS? Can the data from the auxiliary thermocouples and accelerometers be archived in non-volatile in memory for future analysis?

The Chamber

There are a number of items that are important for the proper design of a thermal chamber combined with vibration. Some of these items are required for the proper and safe operation of the system; others are for worker comfort; others can be considered as items which are nice to have but not essential. Consider each of these items and the items previously discussed for your particular application.

Product Accessibility

Does the chamber have a single entry? Is it easy to service the unit under test in terms of attaching cables, making mechanical adjustments, etc., or does it require the operator to use stepladders and bending? Product access needs to be easy so that it does not become a physical hardship to the operator.

Ducting Air

Is it possible to direct air to the product without designing, building or purchasing add-ons or modifications to the chamber? Ducting the air will greatly improve the airflow across the product, thus reducing test times. The chamber should have multiple air discharge ports with balanced airflow and the capability of attaching simple flexible ducting to ensure efficient testing. Two notes of caution:

- Do not restrict the airflow being ducted to the product by leaving the product's outer covers attached—remove them. The airflow within the chamber needs to be continuous and as unrestricted as possible.
- When ducting air to the product, place a thermocouple in the duct's air path. If the thermocouples are only attached to the product, the air temperature may exceed the limits of some of the material in the product.

Work Area Audible Noise Level

Can the operator have a telephone conversation in the room where the system is located without difficulty? Some systems are ineffective in noise reduction. Compare audible noise specifications at *full* vibration levels and at the worse noise level point at the same distance from the chamber (about 1 meter).

Serviceability

Serviceability is another important issue. Is it easy to service the vibrators, when required such as, during preventative maintenance? If the system is used for HASS or HASA, then the stress system down times need to be minimized and access to the vibrators needs to be quick and easy.

Service

Service provided by the manufacturer cannot be overemphasized. Contact customers from the manufacturer's customer list and ask them about the service they have been provided. Is it timely, accurate and will they respond to your needs during non-working hours (if required for HASS)? Contact and visit some of their customers in your immediate area as well.

Maximum System Capabilities

Do the vibration and temperature systems have the capability of producing levels that far exceed what you expect? If the system has more capability than what is needed today, then there is room to accommodate tomorrow's improved products without having to invest in a more powerful system. Do not limit product stress capabilities regardless of what the manufacturer may say.

Post Sales Support

Post sales support can be a critical item especially if one is a novice at HALT and HASS. One needs to ask pointed questions such as, how many HALTs or HASSs have been performed and for whom? What were the results of the stressing? Was the customer able to introduce their product on time or ahead of schedule and failure-free? When was the last time that you did a successful HALT or HASS? The list of questions can go on forever. It is necessary to point out that the bottom line (price) is important. It is also important to remember that you are making a substantial financial commitment and if there is no one at the manufacturer to answer your questions you are left powerless.

Auxiliary Equipment, Operator Safety, and ESD

A properly configured test area requires consideration of the equipment needed to perform testing, fault detection, operator safety, ESD protection, and failure analysis. (Failure analysis requirements will be briefly discussed later in this chapter.)

Each product is unique in its requirements for test equipment and these needs should be carefully considered. The areas that require special consideration include: the interconnection between the unit under evaluation and the test equipment located external to the chamber. Teflon cables are capable of withstanding the temperature extremes that are used in the HALT and HASS process and therefore are recommended over Polyvinyl Chloride (PVC) jacketed cables. PVC, when subjected to temperatures in excess of 90°C, will outgas an extremely toxic and fatal gas (chlorine) and should be avoided. As a reference, PVC jacketed cables and wires are not allowed in building air plenum areas for this reason.

Regarding operator safety, there are a number of items that should be considered so that the operator will not be subjected to any condition that could compromise their health. A list of these items would include:

- The work area needs to be free of elevated levels of nitrogen. This condition can be monitored with oxygen depletion monitors which have audio and visual alarms. (Some monitors have remote outputs that can be connected to the facility alarm system and the LN_2 tank automatic shutoff valve which warn the operator of potentially dangerous conditions.) These monitors should be mounted approximately

30 inches from the floor for effective operation. In the event of a nitrogen leak it will begin displacing oxygen from the floor up because cold nitrogen is heavier than ambient oxygen. Many manufacturers produce quality monitors and you should procure one that meets your needs. Consider the purchase of more than one in the event that one fails or is undergoing preventative maintenance.

- Air exchanges within the work area should meet required codes. If possible, allow an air exchange rate that is higher than usual so that the air within the area is exchanged more frequently.
- If the operator needs to stretch to reach any of the product(s) which are under evaluation on a continuous basis, consider using some type of lifting mechanism so that lower back and leg problems can be reduced. Also, consider a chamber with both front and rear product accesses.
- If the products that are to be handled require an ESD safe environment, consider this, preferably, before installing the chamber. Install a conductive flooring system that meets ESD requirements. If this is not possible, then use wrist straps or an equivalent device. Consult your company ESD coordinator or an ESD consultant for assistance in making the area safe.

Failure Analysis

Many books have been written on failure analysis, therefore the treatment of this subject will be brief but enough for decisions to be made on whether to perform this important requirement.

Individuals who are specialists in the field should staff the lab. They will need to have periodic training to remain knowledgeable regarding new technologies in production as well as those in failure analysis. A budgetary number for the purchase of the equipment to begin a failure analysis lab ranges from $100K to $150K. This estimate includes a stereo zoom microscope but does not include a scanning electron microscope, an acoustical microscope, or the facilities.

If your facility is a low to medium volume facility, then using a local failure analysis lab may meet your needs in terms of quality and failure analysis turnaround. On the other hand, if your facility is a high volume producer, consider investing in an on-site failure analysis lab. It will pay for itself in short order.

Conclusion

Although older vintage equipment can be used for short-term reliability improvements, the dramatic improvements can only occur using the equipment detailed in this chapter. If your budget won't allow for the purchase of a combined system and your product reliability is suffering, then make do with what equipment you have until you can purchase a combined system having very high thermal rates and six degrees of freedom vibration. Don't overlook the other ancillary equipment (programmable power supplies, programmable oscillators, etc.) which will also be required to implement the processes of HALT, HASS, and HASA.

The entire area where the combined equipment will be located needs to be carefully considered as has been detailed in this chapter. One may need to consider heating, ventilating, and cooling (HVAC) as well as power for the equipment and ESD flooring. Much needs to be done before a combined system arrives.

The equipment is only part of the process since access to a failure analysis laboratory is paramount in your quest for high reliability products. The lab must be capable of providing you with the necessary results so that product shipment decisions can be made in a timely fashion.

Six

How to Sell New Concepts to Management

> The man who removes a mountain begins by carrying away a small stone.
>
> **Chinese Proverb**
>
> Great spirits have always encountered violent opposition from mediocre minds.
>
> **Albert Einstein**

Introduction

Many people, including the author, have had to sell new concepts to management. Selling includes technological innovations which we may not be completely familiar with and have very little of the historical information required to gain that knowledge. This chapter will attempt to help the reader become familiar with a reliability improvement program that is radically different from the methods with which he or she may be accustomed. In addition to the contents of this chapter, it is highly recommended that the reader become familiar with *Selling New Reliability Concepts*^G, by Joan Pastor. This chapter will focus more on the technical and financial aspects rather than the human interactions and group dynamics which are critical to a successful implementation.

The material presented in this chapter has been extracted from an actual case study by a manufacturer of consumer electronic products for the time period between 1982 and 1990.

Overview

This chapter will help the reader propose a new concept to management that may be radically different from how they presently assess product reliability. Through the application of this new concept, the products will be more reliable and profitability will dramatically increase. It is assumed that the reader does not have intimate knowledge of the HALT and HASS processes and is now ready to help the organization realize that there is a better and a more practical way of producing very high reliability products. Both processes will need to be reviewed so that an effective presentation can be made.

The Situation Today

Many professionals staff our present product reliability program: statisticians, reliability engineers, product support engineers, test engineers, and technicians. All of them are kept quite "busy." In this example we will look at how many of these people are actually performing productive work and then compare them with those in a different environment, one in which products are mature at introduction. These are the major efforts in our present reliability program:

1. Our present reliability program consists of traditional elevated temperature, steady-state burn-in, product qualification (Qual) testing, and reliability verification testing (RVT). Our field failure rate may not be the lowest in the industry but we are making money on spare parts and field replacement units. We also have a large staff of engineers and technicians who daily answer many phone calls on product specific applications, intermittent product failures experienced by customers, as well as product degradation over time. The two latter issues are the most time-consuming because frequently the engineers don't know what the problem is that the customer is describing to them and subsequently have to discuss the matter with the designer, and do some research and testing on their own.

2. Qual testing is always a bottleneck. The majority of the time is spent trying to make the product meet the specifications in whatever the failing environment may be. Sometimes waivers are written so that the product introduction isn't delayed. Because of the bottleneck in product qualification, our products are typically introduced six to nine months after the projected introduction date—sometimes even later. This, of course, causes havoc in our earnings and growth potential because of the delayed anticipated revenue. Our customers also become upset because we have introduced immature products.

3. During elevated constant temperature burn-in, failures occur at a rate management and engineering feel are acceptable. This, of course, is part of the cost of doing business to weed out some of the infant mortalities so our customers don't see them. This process does ties up all of our day's production. Additionally, a very large area on our production floor has been allocated for burn-in chambers/rooms, carts, controllers, and personnel to manage this effort. The cost of electricity has increased substantially over that which we were paying when this effort began a few years ago. When this effort was initiated, we had many more failures than we are finding today.

4. RVT and reliability growth testing (RGT), which happen just prior to product introduction, are conducted on products at room temperature. Some products are power cycled (to simulate the end-user's application) while others are kept on all of the time. Many of the failures that are detected during this test are explained away. Some failures are believed to be caused by the operator; others are ignored because the end-user wouldn't use the product that way. Isn't the operator representing the actions of a customer and producing defects that would be seen in both worlds? Many other reasons (excuses) are given but these two usually predominate. This test is usually done in tandem with product qualification. RGT and RVT also occupy about one-third of the space needed for burn-in. Incidentally, we have never had a product truly meet the pre-production release reliability goals during our pre-production RVT nor in the field.

Table 6.1 details the staffing for our present program. The figures in Table 6.1 may be conservative but they were the actual numbers prior to the implementation of a new process, which we will talk about shortly.

Table 6.1 Personnel and space requirements under existing program.

Job Function	Quantity	Space (ft^2)
Statisticians	3	300
Reliability Engineers	6	600
Product Support Engineers	5	600
Test Engineers	4	400
Technicians	8	800
Test Support	6	600
RVT/RGT Area		1000
Product Qualification Lab		4000
Burn-in Area		3300
Total	32	11600

Table 6.2 Reliability costs under the existing reliability program.

Category		$, Per Unit	$, Extended
Monthly ship rate (# of units)	1300		
Product selling price		850	1.105M
Average field repair cost		110	
Annual field failure rate	20%		28.6K
Warranty as a % of revenue	2.6%		28.7K
In-house failure rate repair (burn-in)	10%		2.0K
Cost of units for RVT (80 units)		850	68.0K
Cost of units for qualification testing (25 units—one time cost)		850	22.0K
Power cost for burn-in (24 hours × 1300 × #0.06/KW × 30days)		1.44	56.2K

Table 6.2 details all of the expenses that are associated with our present testing methodology. As shown in the table, conservative assumptions will be made when we implement the new process. Incidently, the product for which Table 6.2 is presented was a middle-of-the-road product.

The Proposed Program

The proposed change may be radical and will require a paradigm shift or a new way of conceptualizing reliability assessment. The majority of electronic hardware failures in the 1990s were not due to component failures at all. Most were attributed to interconnects and connectors, product designs, excessive environmental stresses, and improper user handling. In this new environment we will discontinue using many of the reliability processes that were considered to be vital to our business, such as burn-in, reliability growth, calculating reliability numbers, and having a large product support staff to answer problems due to design flaws. With these new concepts we will be able to achieve higher field reliability and improved customer satisfaction. By the way, these "new concepts" have been in use since the early- to mid-1980s.

Let's look at these changes. First, we will begin by step stressing our products (HALT) in a controlled environment long before they're in production.[4] The product should be powered on and performing its intended design function while being monitored during the step stressing process. Step stressing is done for each stress environment until the product's inherent operating limit(s) and destruct limit(s) are reached. At each product limitation, a root cause analysis and corrective action process are performed. At each step the product is allowed to dwell for a minimum of 10 minutes which allows for product stabilization and completion of the full product diagnostic suite.

The stress applied does not necessarily have to be encountered in its end-use environment. We no longer *test to spec* but test to failure. From these failures, one is able to enhance the product reliability through root cause analysis and corrective actions. This, of course, makes the product more tolerant of variations in the manufacturing process as well. For example, a product designed for the office environment might encounter stresses such as extreme temperature dwells and extreme vibration levels that may not be found in its end-use environment. By the time a product reaches the customer, it has possibly been through environments that were not considered possible by its manufacturer. Some other examples include: speed bumps, repeated crevices (expansion joints) on an interstate highway at high speeds, excessive heat during transport, and careless handling during shipment. The customer doesn't care about all of the abuse that the product may encounter on the route to its final destination. The customer wants a product that functions out of the box according to expectations and that doesn't fail during warranty or even after the warranty has expired. If the product fails to meet the customer's expectations, then it will usually be returned for credit towards the purchase of a competitor's product or if it fails before the warranty expires, the customer may likewise choose to get the competitor's product. It's a shame to invest money in advertising to attract a customer and then lose them to the competition, who may not have invested the time and money to attract them in the first place. The picture is clear—produce an exceptional product that far exceeds the customer's expectation. As a result, he or she will seek a replacement product or an upgrade from your corporation—not the competitor. It's far less expensive to retain a customer than it is to attract a new one!

When a failure occurs, it is corrected through a closed-loop corrective action (CA) process.[1] Through the application of the step stress and closed-loop CA processes, a mature product will be introduced to the marketplace. By mature, it is meant:

1. Infant mortalities (failures within the first 90 days of use) will be eradicated (see Figure 1.4 in chapter 1); and

2. Failures during and after a reasonable time following warranty will likewise virtually disappear, (see Figure 1.5 in chapter 1).

Once the product has been made robust through the use of step stressing, combined environmental stresses, and closed loop CA, it is ready for release to production. At this point consider applying an accelerated stress production monitoring process (such as HASS) as your products are ready to be shipped. This monitor will apply repeated stresses (at levels below those applied during the development phase) to either all of the products[4,6] or to a statistical sample.[3] The choice to implement HASS is one that needs to be carefully considered. For this chapter we will assume that all of the produced units will be monitored through HASS.[4,6]

The cost of installing and operating this new process (pre-production and during production) for *one year* is estimated in Table 6.3.

Please note that a more detailed analysis of Tables 6.1 through 6.3 is provided in the sections that follow.

Addressing Potential Management Concerns

Proposing a radical change in production will cause some people to resist. To cope with this resistance, you will need to be prepared to face differences of opinion and conflict. In order to resolve this you will need to be prepared to address the underlying issues such as benefits to each of those involved in the decision process. In other words, demonstrate that this new concept has benefits for each and every person who will be involved in the decision. Remember, as you address the group, you will need to have facts as well as a working knowledge of the concepts so that you can effectively convey them to the group. You will be considered the "expert" by all of those in attendance. It is important to realize that you won't have all the answers and will need to do additional research and reconvene the meeting. Make sure that all issues are recorded and addressed. Your role here is one of consulting and effective negotiating. Allow others to take credit and help with issue resolution but remember that you need to be the "cause champion."

Table 6.3 The operational costs of HALT and HASS for one year with two stress systems.

Function	Quantity	$	Space (ft²)
Stress systems	2	220K	800
Aux. monitoring equip		30K	20
Electricity & LN$_2$		25K	
Facilities preparation		50K	
Cost to Monitor Products	1300	20K	

Assumptions—There are three work shifts stressing 24 products per shift (this equates to a monthly capacity of 1558 units) and that the second stress system is used for pre-production product ruggedization (HALT); 1300 is shown under Quantity column because of inefficiencies in the HASS operation, meetings, preventative maintenance, etc.

In preparation for your meeting, as part of fact-gathering, you may want to consider tabulating the top 10 field failure modes for a current production product. (A Pareto chart is useful for this task.) Armed with this information, arrange to perform step stressing (HALT) on a few new samples of your product so that your field failures can be compared with those uncovered during step stressing. These can be operational field return units. Realize that some of the stressed products may return inoperative but that they can be refurbished and may be used as demos. Experience has shown that the flaw types uncovered in step stressing and in the field will almost perfectly overlay. From evaluating your product, you will be able to address two very important issues:

1. These techniques will work on your product possibly without serious redesigning; and

2. It uncovers the same flaws in far less time than it would take in the field.

These two issues are perhaps the largest hurdles that you will have to overcome.

Addressing the issues of burn-in and RGT/RVT will also require some forethought. Burn-in has worked somewhat in your factory because of the inherent flaws that exist in your product's design, processes, and in your supplier's processes. Burn-in does uncover some of the gross issues. Enhanced product design ruggedization techniques that are performed long before the product is released will assure that eliminating burn-in will be a profit contributor and not a detractor.

In turn, RGT/RVT may become moot points. In the previous reliability effort, we attempted to prove calculations that were based on invalid assumptions. You may recall that this did nothing to improve product reliability.[17] When failures were uncovered during RGT/RVT, some were addressed while most were explained away. Did any of these "explained away" failures ever occur later in the field? Eliminating RGT/RVT as an unnecessary expense, not unlike burn-in, will increase your profitability. As you can see, we are saving money in one sense and generating money in another. How? By replacing the *old* methods with a simplified and foolproof process (HALT).

The generation of revenue is another area that you will want to consider in adopting these concepts. Some thoughts could be:

- *Earlier product introduction.* Earlier product introduction creates early revenue due to shorter development cycles. Since the new products are getting to the market before your competition, you can possibly set a higher initial selling price and lower it at the appropriate time so that you become the pacesetter.[10] Delays in product qualification are now minimized if not totally eliminated. Potential areas for bottleneck may include regulatory compliance but not the environmental stressing portion of product qualification. See Figures 1.1 through 1.3 in chapter 1.

- *Mature product and introduction.* At introduction, your product will have few, if any, field failures. This has a two-pronged benefit. First, your warranty costs, repair costs, and warehouse space for replacement parts will dramatically decrease. Second, your in-house rework time will decrease. Since you have ample design margins, you can tolerate shifts in processes that once caused production shut downs and rework. See Figure 1.5 in chapter 1.

- *Increased customer satisfaction.* It has been found that every satisfied customer will tell three acquaintances, while the dissatisfied one will tell eight to sixteen.[16] Bad news spreads fast while the good news takes a little longer. As you integrate these concepts and begin to monitor (if you're not already doing so) customer satisfaction, you will see a marked improvement in the satisfaction index.

- *Overhead reduction and reassignment.* All of the space and people required in the old program can be reassigned to roles that enhance profitability. The space for burn-in and RGT/RVT can now be used for producing more products or developing new ones. Statisticians now can work on statistical process control and assist with design of experiments (DOE) to make new designs more robust. Reliability engineers can work on adopting new concepts early during the product's design phase.

- *Fewer units are required for pre-production evaluation.* You will no longer have to produce so many units for the obsolete tests. Instead you can produce them for shipment and revenue. The new paradigm requires only a few units. Possibly, you may want to consider beta testing of your product prior to production release if you are in the commercial market since these tests can provide extremely valuable information.

The Savings

In addition to the five items listed above, we have some figures from which we can calculate the savings. The HALT process can be easy to justify since many have written about their successes. This justification will be dependant on your product, whether it's a consumer electronic product, is a product for the telecommunication industry, for the automotive or whether it's for the aerospace industry, many good papers exist from which to assist you with the justification. HASS and HASA papers are not as prevalent. Some have justified HASS on the basis as it being an insurance policy or safeguard against shipping defective product. Regardless of which process that you are trying to justify, you will have to do some research. You may want to begin by thoroughly researching the failure modes of one of your existing products and estimating the failure rate for its replacement. Sometimes the new product may have the same failure rate as its predecessor but its increased shipment volume may overwhelm the repair center. Is this situation acceptable to management? IBM (Sequent Computers) found that the defect level of the new computer without HALT and HASS would be unacceptable because of its added complexity. They concluded that using HALT for the design phase and HASS during production could substantially reduce their defect levels. After justifying the purchase of two systems (one for R&D and HALT; the other for production and HASS), they compared their estimates to what actually happened. They found:

- A better than 100 percent improvement in boards that were now defect-free in production (a HALT contribution by removing design defects and adding ample margins);
- An 82 percent reduction in board build cycle times (both HALT and HASS contributions); and
- The failure rates on many components were essentially *zero* in production (a HALT contribution).

All of these contributions exceeded their expectations![6]

For the sake of simplicity, we will add on to the previously shown tables to show the justifications for new techniques.

Table 6.4 shows a comparison of space and staffing requirements:
From Table 6.4, it can be seen that with the new processes, the total square footage is reduced by 65 percent and the total personnel required by 63 percent! The staffing from the present system would be reassigned under the proposed system scenario. For instance, the reliability engineers, even though the head count remained the same, would be assigned to work with the R&D team throughout the project as a team member to help with reliability issues. Reliability engineers would not spend time calculating reliability (MTBF) numbers. Derating and design review would become their fortes along with HALT and HASS. On the other hand, the technicians, test engineers, and test support staff could be reassigned to more productive assignments such as: R&D (for the engineers), production assembly for the test support staff, and qualification or product safety endeavors for the technicians. The old equipment (burn-in chambers and test setups), for the most part, would not be needed with the proposed system.

A note on MTBF calculations. Some industries require MTBF numbers for their customers. Realize that as you implement HALT (and possibly HASS) that your field failure rates will dramatically decrease over time. Once the failure rates begin to indicate a downward trend, the λ values in the component database for the MTBF program will have to be adjusted downward as well. The result of this is a dramatically improved real MTBF!

Table 6.4 Comparisons of personnel and space requirements for present and proposed reliability programs.

Job Function	Present Program Quantity	Present Program Space (ft²)	Proposed Program Quantity	Proposed Program Space (ft²)
Statisticians	3	300	3	300
Reliability Engineers	6	600	6	600
Product Support Engineers	5	600	2	240
Test Engineers	4	400	0	0
Technicians	8	800	1	100
Test Support	6	600	0	0
RVT/RGT Area		1000	0	0
Product Qualification Lab		4000		2000
Burn-in Area		3300		0
Total	32	11600	12	3240

Table 6.5 Comparisons of on-going costs for present and proposed reliability programs.

Category	$, Present	$, Proposed
Product selling price ($850)		
Monthly ship rate (1300 units)		
Avg. field repair cost ($110)		
In-house failure rate repair at burn-in (10%)	2.0K	0K
Field failure rate (20% vs. 1%)	28.7K	0.1K
Cost of 80 units for RVT	68.0K	0K
Cost of units for qualification testing (25 units)	22.0K	22.0K
Cost per KW electricity ($0.06/KWh)		
Monthly cost for 24 hour burn-in	56.2K	0K
Total	176.9K	22.1K

*Please note that an assumption made in Table 6.5 is $15.00/hr for in-house repair labor rate.

Table 6.5 shows a comparison for the costs of testing.

The cost of testing under the two systems also shows a dramatic decrease of 87 percent! Most of this reduction is due to the fact that old programs, which were ineffective, are being discontinued and the new field failure rate will decrease substantially. This can be further decreased if some of

Table 6.6 Costs for proposed reliability program.

Function	Quantity	One Time Expense	On-going Expense	Space (ft²)
Stress systems	2	220K	36K	840
Aux. monitoring equip*		30K		20
Electricity & LN$_2$			25K	
Training	1	7K		
Facilities preparation		50K		
Total		297K	61K	860

Assumptions for Table 6.6:
1. There are three work shifts stressing 24 products per shift (this equates to a monthly capacity of 1558 units).
2. The total equipment cost could be higher if new equipment is utilized but the assumption made is that some existing equipment will be used and some ($30K) would be a new purchase.
3. Use 5-year Sum of the Years Digits (SOYD) for the stress systems depreciation (first year $3k per month and 5th year $611 per month). You may use linear depreciation. If so, change this value accordingly.

Table 6.7 Summary of savings between programs.

Parameter	Old	New	Reduction %
Personnel	32	12	63
Space (ft²)	11,600	4100	65
Test cost	$176.9K	$22.1K	87

the Qual testing is eliminated because of the extended stresses of HALT. Some of the Qual test equipment can and should be used in the new paradigm but only after the product has been ruggedized through HALT.

The cost of installing and operating this new process for one year is estimated in Table 6.6.

Table 6.6 indicates that the new program will require approximately 420 square feet per chamber. There will be a one-time expense for installation and training (in the new techniques) of $297K and an on-going expense of $61K per year. This increased expense is small when compared to the reduction in space and staffing when going from the present system to the new system.

Finally, Table 6.7 summarizes the effects that the new paradigm will have on the facility. The conclusions are very clear. The new paradigm is a win-win situation. Personnel requirements, space allocation on test costs have all decreased dramatically. The only items remaining are to estimate the return on investment and the increased profitability.

Conclusion

It has been clearly shown that the proposed reliability improvement program is far less costly and has a much higher effectiveness than a reliability program using traditional methods. The new program requires far less space and its overall implementation costs are much less than those used by the dated, ineffective programs. A cost that should be common to either program is that of failure analysis and therefore, its costs are not included.

The new enhanced reliability program has other features which one cannot immediately put a monetary value on, such as increased customer satisfaction, increased market share, reduced costs associated with warranty administration, an increased warranty period (some have actually shown that the total warranty cost actually *decreased* even after extending their warranty period coverage), far less factory rework and scrap, enhanced productivity, and less inventory. All of these are expenses that detract from the bottom line profit. Only those items that could be measured were shown in the previous calculations.

As one begins down the new reliability improvement road, HALT obviously needs to be done. The entire process needs to be religiously followed. As a result of this endeavor, a product will be introduced that is robust with ample design margins and very little, if any, increased cost over the old design. Care must be taken to provide coverage of the production processes in the event that they go out of control and the change(s) is not readily detectable under normal production controls. Traditional testing methods will not detect these deviations and an accelerated stress technique must be applied here as well. This monitoring process can either screen the entire production (HASS) or perform a statistical audit (HASA). Each process has its place. HASS needs to be used on processes that are not within statistical control (or contractual requirements) while HASA can only be used when the production processes are in statistical control. HASA utilizes all of the techniques that are used in HASS but it also requires some statistics. For even the novice, the statistical equations have all been selected and examples detailed in order to make its implementation easier.

Seven

Some Commonly Asked Questions and Observations

> Everything should be made as simple as possible, but not one bit simpler.
>
> Albert Einstein

How Would Someone Compare ESS and HASS?

In the simplest terms, a test is either passed or failed. The objective of a test is to ship all of the samples that have passed the test to the end-user. In most situations, units that fail the test are made to pass the test by some form of repair or possible modification. Environmental Stress Screening (ESS) is a stress or combination of stresses that when applied to a product, uncovers latent defects. These defects would, under normal production test conditions, remain undetected until the product has been in use for some time in the field. ESS is far less demanding on a product than other forms of design marginality product stressing like HASS. In other words, ESS is conducted with stresses that are usually below what the product could actually sustain but because of cultural or other biases, the product is usually not taken "beyond spec." ESS is applied to the product during the production phase of its life cycle typically without prior margin determination and design ruggedization for that product. (See chapter 1 on HALT for more details on product ruggedization.) HASS, on the other hand, assumes that the product has been ruggedized prior to applying HASS. If prior product ruggedization has not been done, HASS will uncover both the production issues and most of the design issues. This philosophy, of course, is counterproductive because the cost to correct a design deficiency in production is very high and time consuming. Design deficiencies need to be corrected during the product design phase. Had stress levels beyond the traditional ESS been used (as in HASS) and had preproduction product marginality and robustness been performed (as in HALT), the production product screen would be solely used to monitor production processes and not to discover design deficiencies. This is a major philosophical difference between the traditionalists and those who use accelerated stress techniques on their products from the product's earliest stages and throughout its production life. See Table 7.1 for additional clarification.

What is HALT in a Few Words?

As previously mentioned, HALT is a process by which a new product is evaluated in terms of its operating and destruct margins through the application of stresses. These stresses are applied to the product while it is performing its intended design function, for example, a disc drive would be writing, reading and doing random accesses while being monitored for its specifications. Although these applied stresses may not be encountered by the product in its end-use environment, the applied stresses cause the defects which would usually takes weeks or months to surface in the field, to be uncovered in a just few minutes or hours. HALT is commonly used to eliminate design problems and infant mortalities in a product before its market introduction. This, of course, will eliminate the design's weak links, process problems, and

Table 7.1 A comparison of ESS, HALT, and HASS.

	Test	Stress Beyond Spec	Prior Product Ruggedization Required?	When Used?	Design Defect Discovery?	Process Defect Discovery?
ESS	Yes	Usually not	No	Mfg.	Some	Some
HASS	Process	Yes	Yes	Mfg.	Some, but not its focus	Yes
HALT	Process	Yes, very	NA	Design	Yes	Yes

the product's package issues. Performing far beyond their published specifications, the products could have overall warranty expenses that are lower than their competitors and leave their owners happy with the product. Other factors that are not commonly considered or recognized up front that can also be enjoyed by the producer are higher profits, earlier market introduction, and less engineering required for product support. Finally, the new product should be capable (with a marketing focus) of obtaining an increased market share.[10]

How Would One Compare Product Qualification Methods and HALT?

In order to produce a product that will exceed the customer's expectations and leave them delighted, techniques beyond the traditional methods must be employed. These techniques have been in use since the early 1980s by some companies that have been enjoying tremendous success as a result. Applying stresses to a product that are "beyond the specs" has been their motto. Although foreign to some at first, these techniques really do work well. Any stresses can be applied within reason. For instance, a product that is designed for the office environment dictates price-consciousness on the part of the end-user and will be used in a fairly benign environment, may have published specifications for temperatures from 10°C to 40°C. During production stressing, this product could be subjected to a temperature range from 75°C to -80°C with other stresses being applied as well.[3] Some questions come to mind about this product's transportation from the manufacturer to the retail store and user. Before reaching the end-user the product may have been subjected to the heat of the Mojave Desert in the enclosure of a truck (>140°F) as well as the periodic discontinuities of a concrete paved road. At the receiving dock it is then thrown, and not gently placed, on a pallet. Obviously, the product has experienced stresses that are far beyond what typical product qualifications may require. If this were not the case, then products would not be failing in the field and the reliability community would not be searching for better methods to evaluate and ruggedize their products.

Let's look at the HALT stresses that may be applied to this product during its product development phase. It would not be unrealistic for the temperature range to be from -80°C to +100°C with 10-minute dwells at each of these steps and a product temperature change rate of at least 30°C per minute. Vibration step stressing and dwelling could be another stress that would be applied in the range of 15Grms to 20Grms using six degrees of freedom vibration. This vibration can be best visualized as the three orthogonal vectors (x,y, and z) with three rotational vectors about each of these, all occuring simultaneously. This type of vibration has many benefits of which throughput and synergism are at the top of the list. Most motion that a product experiences in use does not typically originate from a source which has a single degree of freedom but has multiple degrees of freedom. So why not use technology which more closely approximates the *real world* which is not severely restricted by some controller but is random in its occurrence?

Our evaluation of the product should not be limited to just temperature and vibration but other stresses as well, such as power cycling, voltage margining, clock margining, and any others which can be imagined. Remember that after applying each of these stresses independently, they all must be applied concurrently.

Is HALT for Quality Improvement or Is It Intended to Replace RGT and MTBF Tests?

HALT's main impetus is for product reliability improvement. Quality is also improved. There is one main difference between quality and reliability: Quality is used to measure the performance of a product when it first arrives to a customer; reliability is used to measure performance and longevity. With this in mind, HALT is more adapted to reliability than quality. Once HALT is completed, HASS should be implemented. Once again, HASS is a very effective tool for on-going quality improvement and monitoring.

Since the goals of RGT and MTBF are for the measurement of product reliability and the goal for HALT is to improve the product reliability, they are not replacements for one another. Presently, HALT does not allow one to calculate a reliability number. One needs to be cognizant that the assumptions for reliability numbers are inaccurate. In contrast to HALT, which uncovers real defects and thus leads to real product improvements, one would want to correct these weaknesses so that the product reliability is increased. If RGT and MTBF tests are required, the recommendation is to perform a HALT first on the product and then perform the reliability tests. As a result, the reliability tests will be shortened and very few, if any, failures will be found.

Is There Any Merit to Subjecting a Product to Stresses Far Beyond Its Design Specifications?

Product specifications should not be the controlling item in determining the levels of product stressing during HALT or HASS. It has been shown that the failures encountered during HALT *are* relevant and will cause field failures if not corrected. To prove that this is the case, subject a product that has been in the field (and has a Pareto of field issues) to HALT without its corrective action in place. After concluding HALT, compare the HALT results with the Pareto of field failures. This has been done many times and the failures are amazingly the same, almost without exception. In other words, failures encountered during HALT will become field failures almost without exception. We must stress the product beyond its published specifications during HALT to ensure that the tails of the margin distributions (operating and destruct) will never come close to the product design (published) limits. Any time that the tails of these distributions touch the published limits, failures occur.

It is interesting to see how much a product can really withstand. In other words, design the product, subject it to the stresses and let the accelerated stress program tell you what the product weaknesses are, understand them, and then correct them.

HALT's objective is to provide a robust product so that the product's strength distribution will never be encroached by any stress distribution. This phenomena normally occurs as a product ages but the reliability boundary can be maintained throughout the product's life. The product strength distribution can be narrow (no product ruggedness) or very wide (a very robust product). For most products, the designer can not control the product's end-use environment and this end-use environment will apply various stresses to the product. For instance, in an office environment, the product may be inadvertently subjected to a sudden rush of air as a door is opened or a temperature change caused by air conditioning and heating system or to the sudden dropping of a heavy object near the product. These stresses, although possibly innocuous, may cause the product with a narrow strength distribution to fail. The product with the wider strength distribution may not fail. Over time, the product's strength distribution will degrade and the ruggedness (strength distribution) of the product will be the primary factor in the determination of whether it fails prematurely or not at all.

What Are Product Specific and Generic Stresses?

There are two types of stresses that should be applied to a product—product specific and generic. Product specific stresses are those which are known to accelerate failure modes within a given product. For example, in an image scanner, a low DC voltage in combination with high oscillator frequency may cause the product to behave abnormally. Generic stresses are usually considered to be six degrees of freedom vibration, high temperature product ramp rates ($\geq 30°C$ per minute), wide temperature range ($\geq 100°C$), and short (usually 10 minute) temperature dwells.

If HALT and HASS Are So Great, Why Isn't Everyone Using HALT and HASS?

Although the techniques have been around since 1980, many manufacturers have been required to conform to pre-described specifications and product qualification methods. This, of course, may limit the creativity of the engineers because they will design to what is expected in order to make the product qualify. Engineering schools still require calculations using the Arrhenius equation. If the world dealt with chemical or time dependent failures exclusively, then this teaching would be justified. Many types of stresses (such as vibration, rate of temperature change, power cycling, etc.) may exacerbate today's component failures.

A "shortcoming" that HALT and HASS both have is that they haven't been totally quantified by equations and theories. Recently, work has begun and papers have been written by practitioners trying to quantify these processes. [2, 21] Some of the issues that are being addressed include: why a sample size of four can uncover field failure rates that are far less than 1 percent and how to calculate product reliability improvements from one HALT to the next. [21]

Interestingly enough, the manner in which ESS experts have been applying the Arrhenius equation has been incorrect.[11] Still another point is that using MIL STD-217 yields results that are not consistent with field data. MIL STD-217 has undergone many revisions (at least six) to try to make its computations agree with field data. Many studies show inconsistencies between calculated and field results. If the data has had such a disparity for so many years, why use it? In a Boeing Commercial Airplane study,[1] they found that the actual and calculated MTBF ratios varied from 5.26 to 0.64. This meant that the actual product reliability could be as much as 5.26 better than that which was calculated or only 0.64 as good.[17] By the way, the same manufacturer under the same process controls, parts suppliers and MIL STD-217 criteria built all of the units that were used in this study. Some have been successful in changing the program's failure values to reflect what their product experience in the field and these calculations have provided more accurate MTBF estimates than the standard 217.

Since 1990, some users of these techniques have been willing to share their product and business improvements, but this number is relatively small. The majority has not had the desire or time to share these techniques because of the competitive advantages that techniques can realize, and hence, they don't want their competitors to know why they're *pulling away from the pack*. Also, time is a luxury that many can ill-afford to invest in publishing their findings.

Is One HALT Enough?

In the author's experience, so much is usually found in a HALT that some redesign and limited retest are usually in order. The second test may (and usually does) uncover more problems and so a second redesign and/or process iteration may be required. Usually, these "redesigns" are small changes. The retest on redesigned hardware should continue until the test results are satisfactory and generous margins are attained. If root cause analysis, corrective action, and failure analysis are quick, the time between HALTs can be minimized and the entire HALT process can be accomplished more quickly than traditional, ineffective qualification methods.

After the product has been ruggedized through HALT and released to production, it is recommended that periodically (usually every six to nine months), HALT be repeated to verify that none of the margins have degraded. This degradation, unless substantial, may go undetected in HASS because of its own reduced stress limits. Therefore, it is imperative that one repeat HALT periodically. These degradations could be subtle changes in the components that may not exhibit problems during HASS (because of the lower stress levels) but will certainly manifest themselves during HALT.

At What Product Level Should HALT Be Performed?

HALT can and should be done at each assembly level including:

- Circuit board;
- Subassembly, such as, a power supply or card cage;
- Box or rack as a subsystem; and
- Total system level.

The rationale is two-fold. First, each circuit board and sub-assembly needs to be evaluated through HALT so that optimum ruggedization is achieved (ample margins). Second, at each added level of product complexity, there is possible reduction of stresses (both vibration and temperature) to the product because of added product weight, permanently attached product covers, etc. Fixtures to secure the product at higher levels of assembly to the vibration table also become more complex and heavier. Having stated this, abnormal product conditions not measurable at the lower level (board) HALT may manifest themselves only when configured in the final product. Examples of this would be clock delays, signal degradation, connector robustness, etc. Therefore, consider performing both a board level and product level HALT.

Who Should Be Involved with the Accelerated Reliability Program?

Accelerated reliability programs contain complex tasks requiring many talents and skills. As a result, the following major functional areas are usually included:

- *R&D or design.* Their designs are the ones that we will be evaluating and they should be side by side with the HALT engineer when HALTs are being performed.

- *Test engineering.* This group may be involved with the diagnostic development as well as production testing of the product. They and/or the designer should fully participate in the HALT.

- *Advanced manufacturing.* Changes made in HALT can have significant consequences on this group as well as the production group. This group is usually involved with process automation.

- *New product introduction and manufacturing.* This group is usually involved with the new design and its documentation as well as the manual assembly process.

- *Materials engineering.* This group is also known as procurement engineering or supply line engineering, and may be responsible for the supplier selection and they may be the interface between your facility and the supplier when the accelerated failures begin to surface.

- *Quality and reliability.* Usually, but not always, this group is the facility "champion" for accelerated stressing. Over time, the process may be permanently assigned to production (HASS and HASA) and to R&D (HALT).

- *Program management.* Program management usually does not actively participate, but must have some understanding of the process or your program will lack support from them.

It is strongly recommended that a standing committee be formed with a representative from each area. This committee should meet regularly to discuss open issues, progress, and decide on issue ownership. This committee can be discontinued once HALT and HASS have been integrated into the culture.

An alternative method follows:

- Schedule the HALT and invite the designer(s) and test engineers to attend the HALT. They will prove to be invaluable in root cause diagnosis.
- Complete the HALT report and distribute it to the designer(s) and managers.
- About one week following the issuing of the HALT report, schedule a post-HALT meeting. At this meeting have the test engineer, the HALT engineer, the designer and anyone who could help drive the issues uncovered during the HALT to understand root cause and to implement corrective action in a timely fashion.

Physically, Where Should HALT Be Performed?

Ideally, HALT should be done at your facility because you will have access to the experts when a problem occurs. The next best place to perform a HALT would be at a contract laboratory that has the expertise to do so. Realize that not only will the product that is to be evaluated have to be transported to the facility but the necessary test equipment as well. The lab needs to be staffed with HALT experts who demonstrate sound engineering judgment. When doing the HALT at a remote site you will need to have the product designer present. The next option is to consider leasing the equipment and contracting an expert to assist you in your first HALT(s) at your facility.

How Many Units Are Required and What Can Be Done with Them Once We're Finished with HALT?

During early prototype fabrication, few units will be available. However, the search for problems will be fruitful. When the identified problems have been fixed, it is time to repeat HALT. There will be fewer problems found at this stage. Deciding how many times to test may be affected by what we do with the units afterward. In other words, are they disposable? Some units that have been subjected to HALT can be refurbished and used as demos or beta test units. Others can be refurbished and used as Proof Of Screen (POS) units if HASS is to be done. Regardless of the number, one should perform HALT after *every* product modification. Once in production, re-HALT every six to nine months.

Another consideration is that the cost of some prototypes can be several thousands of dollars. Obviously, we don't want to destroy them and there is no need to do so. Remember, that we are not truly destroying the product when we reach the Destruct Limit (DL), we are only finding a limit. Granted, the product will not recover when the stress is removed and in order to have a functional product again troubleshooting and repair will be needed.

Through many HALTs, the author has found that a sample size of four units (with a fifth as a spare) for the HALT ensures that all of the product's failure modes will be found.

Why a Cultural Change May Be Required in Order to Perform a Successful HALT

The expectation to uncover problems is required in the accelerated stressing program, whereas many are accustomed to *passing the test*. This philosophy is still prevalent and can mean that one carefully hand-built unit is run through the qualification tests with the goal of passing. Once one unit has passed, the design is frozen and production is begun. This approach is certainly different than HALT where everything imaginable is done in order to force failures. A few *misguided* quotations from experience and a paper,[1] may illuminate the points to be made:

"There can't be a problem, we've always done it this way!"

"But it only happens below -20° C!" (spec was -10° C)

"No problem, it only happened in one out of ten!"

"It's only a random failure!"

"It's just an infant mortality!"

"Don't worry, it's only a process problem!"

"No problem, we'll fix that after production release!"

"Well, it wasn't a hard failure!"

"Of course it broke, it wasn't designed for that!"

"Where will this product ever see that kind of stress?"

"Of course it failed, you took it over spec!"

"The components are rated for 0°C to 70°C. Naturally, these stress levels will cause the product to fail."

The last four are some of the most difficult misconceptions to overcome. Many individuals think that anything "over spec" is unreal, unwarranted; and just plain unrealistic! Experience with proper HALT and HASS techniques over many years has shown that stresses up to the fundamental limit of the technology will produce relevant failures. The proper application of the techniques requires a positive attitude to improve the product to the fundamental limit of the technology in any. One must discard the philosophy of "If it is not broken then don't fix it!" and replace it with "If it is not perfect, then we'll continue to make it better!"

Figure 7.1 shows the results of a survey of some manufacturers that perform HALT. The results clearly show that the designer could and should use 0°C to 70°C parts and let HALT indicate whether the product has a weakness or not. Be careful not to jump to conclusions when a failure occurs—it may not be due to the part but may be due to the improper component selection (value) of another part, software, or test methods. Another point worth mentioning is that going beyond the published component or product specifications does not mean the product will be destroyed. It may mean that the product will not meet all of its published specifications.

Can a Conventional Chamber and Vibration Table Be Used to Perform a HALT?

HALT can be performed on any type of equipment. It is necessary to push the product's performance to stress limits which are not possible with equipment that are not designed for HALT and HASS. Among these stresses are: temperature range, temperature ramp rates, six degrees of freedom vibration, and the capability to combine temperature and vibration stresses.

Since the higher product thermal ramp rate systems are capable of outperforming the standard thermal chambers by greater than six to one (≥60°C per minute versus 10°C per minute), the user can uncover the product defects in one-sixth the time. Additionally, the thermal range is limited in conventional chambers that will inhibit flaw discovery.

The vibration system has six degrees of freedom vibration. That is to say, three translational and three rotational axis. With this type of vibration, all frequencies (<100Hz to >2KHz) are always present thus, precipitating flaws which usually go undetected with standard vibration equipment. The frequency range for the vibration is just as important for high-mass and low-mass components. Vibration systems that have extended low and high frequency vibration levels are not existent in all six degree of freedom vibration systems.

124　Chapter Seven

Should Commercial or Industrial Rated IC's Be Used?

Company	State	Product	# ICs	Com/Indus	# Passive	Temp−	Temp+	Vib	Comments
3Com	IL	16 Chnl A &D Modem	450	Com	1800	−65	105	45	Technological limits
AlliedSignal	WA	Ground Proximity Indic	75	Both	500	−85	130	35	Mixed due to parts availability
AlliedSignal	KS	Flight Display	500	Com	1500	−95	105	45	LCD display blanks
AlliedSignal	AZ	Fuel Cntl Computer	220	Ind+Mil	1200	−70	130	35	6 boards. Use MTO for cost. Mtd in cargo bay. 100, 8x10" boards. Designer's choice.
Boeing	WA	777 Avionics	75	Both	450	−60	150	25	Application specific
EMC	MA	Switching PS	75	Com	300	−40	100	17	Hot limitation due to caps
EMC	MA	Host Controller	100	Com	500	−40	80	15	SRAM charge hold
Hewlett-Packard	WA	DeskJet printer	12	Com	70	−60	100	8	Low vib due to connector
Hughes	IN	Military two-way radio	131	Both	2000	−65	105	50	Parts availability
Intel	OR	Large server	250	Com	3000	−45	95	30	BGA's, multiple voltages
MCDATA	CO	Fiber Optic Data Sw	3000	Com	15000	−90	100	30	New iteration, even better
Nortel	Ott	FO Trans Ana/Dig Sys	100	Com	700	−45	75	15	Technological limits
Ohmeda Medical	CO	Oximeter	100	Com	600	−55	127	47	Published spec is 0 to 55C
Otis Elevator	CT	Elevator controller	50	Com	100	−30	120	35	MOVs break, FETs, timing at low T
Sequent	OR	Lynx card	120	Com	2000	−55	85	45	Controller failed-not lynx
Sony	CA	Display monitor	50	Com	200	−40	120	40	They pursue destruct after HALT
Storage Tek	CO	Tape Library	50	Com	150	−60	85	25	Motor limitation
Tandem	CA	200MHz Fault Tol CPU	100	Com	1000	−40	80	15	Memory and CPU
Tektronix	OR	17" Hi Res Color Display	40	Com	300	−60	85	30	Plastic melted
Tektronix	OR	Network Computer	50	Com	200	−100	145	50	10-3W ICs; PPM from 3500 (1st gen) pre HALT to 60
Teradyne	CA	1.2GHz Dig PCA	1000	Com	7000	−40	85	20	(2nd gen) w/HALT
Textron	MA	Missile Sensor Fuse	145	Com	400	−40	125	30	Mech parts low temp 15 Boards, 40VLSI & 5 RF Hybrids
Worst Commercial Limits						−30	120	35	
Best Commercial Limits						−100	145	50	
Mean of Commercial Limits						−56	101	30	
Worst Other Limits						−60	150	25	
Best Other Limits						−85	130	35	
Mean of Industrial Limits						−70	129	36	
Mean for All Products						**−59**	**105**	**31**	

Figure 7.1 HALT users survey.

Are All Six Degrees of Freedom Shakers the Same?

No. There are at least two basic designs (patented), one uses a rigid vibration table while the other is segmented. The vibrators should have the ability to "smear" the line spectra so that the modes of the product can be excited and an unrealistic mode is not excited. The two basic techniques (again, both patented) for spectra smearing can be accomplished through piston rotation as it moves through the cylinder from end to end or stroke changes through a modulator valve system.

The bandwidth of the input spectra is also important because some tables have frequency content from below 10Hz to about 10KHz, while others are limited from about 200Hz to 2KHz. The wider frequency content can be imperative to excite large components at low frequency as well as to excite at higher frequencies, the small SMT devices that are widely used in today's production.

Are There Any Known Problems in Applying Product Temperature Ramp Rates of ≥60°C per Minute?

This issue has been discussed with IC designers and they have stated that there was no reason why 90°C per minute could not be used. There has been a report of ASICs Application Specific Integrated Circuits (ASICs) which were limited to 20°C per minute during the product's early development phase but were subsequently able to operate to 60°C per minute by production release.

Are There Any Advantages to Performing Sequential Rather than Combined Stress Regimens?

None. This method (sequential) has been in use for so long and its rationale stems from the lack of combined systems years ago. Many of the early papers (1970–1985) demonstrate that one or the other as the "best" sequence. With today's combined thermal and vibration systems this issue is a moot point.

Do HALT and HASS Just Uncover Electronic Defects?

Absolutely not. Over the years, HALT and HASS have been applied to not only electronic assemblies but to electromechanical assemblies with equal effectiveness. Gyros, disc drives, motors, pumps, valves are a few assemblies that have been evaluated and improved by the processes. These techniques have also been applied to mechanical devices such as printer mechanisms, disk drive mechanisms, plastic lawn sprinkler heads, automotive window mechanics, etc.

Can HASS Eliminate My Production Steady State, Elevated Temperature Burn-in?

The most direct answer is yes. For those industries that are regulated by federal agencies, this may require a bit of effort. If careful data has been accumulated on failures from the burn-in process, then one could step-stress a unit (in order to find its operational limits only) and perform HASS on all production units without ruggedizing the product. This would then allow for an equal comparison of the ESS and HASS results. After this data has been presented to and approved by the agency, a rigorous HALT should be performed to find all of the design deficiencies and correct them so that HASS only uncovers process defects. This process of getting agency approval may be simplified as stated here, but the methodology has merit.

Burn-in at the final product assembly level is not very effective today because it typically is conducted at room temperature or at a slightly elevated (*100°F*) temperature. If HALT has been performed and component failure rates are minuscule, very few if any, failures will be found. It is interesting to note that most integrated circuit manufacturers perform an extended, elevated temperature burn-in on their components at 150°C for days.

At What Levels of Temperature and Vibration Can I Consider the Product Robust?

Although the answer is somewhat specific to product application, generally the following operation ranges have been observed on many product HALTs:

- Temperature range: -70°C to +100°C;
- Temperature rate: 60°C per minute; and
- Vibration: Up to 30Grms measured on the product.

Please note that these are printed board level stresses and they are not a requirement. Temperature and vibration dwells for HALT require a minimum of 10 minutes each. During the production screen (HASS), these are the two basic stresses that can be used. (Please refer to chapter 2 for additional details regarding precipitation and detection screens.)

Prior to implementing the HALT process, you may want to propose the following concept to the design team: any failures found during HALT within the range as depicted below will have root cause and corrective action implemented. This would also include combinations of these stresses. These are product response levels. (See chapter 1 for more details on this proposal.)

- Temperature: -70°C to +90°C at 60°C per minute
- Vibration: 28Grms

Any failures found beyond these limits will have a root cause and corrective action identified, along with a project impact analysis completed. With this information a reliability assessment team is to be convened to discuss the viability of implementing these changes. The decision to implement the changes rests with the group and not one person who may or may not make the *best* decision for the product.

How Can You Justify Doing HALT on Products with a Very Short Field Life?

With products that have a short expected field life, the validity of HALT still prevails. Regardless of the expected product field life, all products have some failure distribution and one must be concerned that the product's failure distribution tail does not enter its life window. The warranty window should not be a concern when determining the product life for a short-life product. The warranty should be at least as long as the expected field life so that customer satisfaction and loyalty are maintained. In other words, the product should not fail until long after its warranty has expired. Also, don't overlook the shipment and end-use abuse that the product will be subjected to during its lifetime. All of these issues can be uncovered during HALT.

Appendix A
The Derivation of Equation 3.1

The derivation of equation 3.1 in chapter 3 follows.

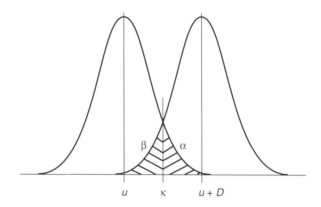

Figure A.1 Decision Criteria.

Let $\mu = p$ and $\sigma^2 = p(1-p)$

$$Z_\alpha = \frac{k - \mu}{\frac{\sigma}{\sqrt{n}}} = \frac{\sqrt{n}}{\sigma}(k - \mu)$$

$$n = \frac{Z_\alpha^2 \sigma^2}{(k - \mu)^2} \; or$$

$$k = \frac{Z_\alpha \sigma}{\sqrt{n}} + \mu \text{ also,}$$

$$k = \frac{Z_\alpha \sqrt{p(1 - p)}}{\sqrt{n}} + p \qquad \text{Eq A.1}$$

$$Z_\beta = \frac{(\mu + D) - k}{\frac{\sigma}{\sqrt{n}}} = \sqrt{n}\,\frac{[\mu + D - k]}{\sigma} \text{ and,}$$

$$k = \frac{-Z_\beta \sigma}{\sqrt{n}} + \mu + D$$

$$k = (\mu + D) - \frac{Z_\beta \sigma}{\sqrt{n}} \qquad \text{Eq A.2}$$

By substituting from Equation A.2 into Equation A.1 and solving for n,

$$n = \frac{Z_\alpha^{\,2}\sigma^2}{\left[\dfrac{-Z_\beta \sigma}{\sqrt{n}} + \mu + D - \mu\right]^2}$$

$$Z_\alpha \sigma = \sqrt{n}\left[\frac{-Z_\beta \sigma}{\sqrt{n}} + D\right]$$

$$Z_\alpha \sigma = -Z_\beta \sigma + \sqrt{n} \times D$$

$$Z_\alpha \sigma + Z_\beta \sigma = \sqrt{n} \times D$$

$$\sqrt{n} = \frac{(Z_\alpha + Z_\beta)\sigma}{D}$$

$$n = \frac{(Z_\alpha + Z_\beta)^2 \sigma^2}{D^2}$$

Substituting $p(1-p)$ for σ^2,

$$n = \frac{(Z_\alpha + Z_\beta)^2\, p(1 - p)}{D^2}$$

Glossary
Terms and Acronyms

If you bet on a horse, that's gambling. If you bet you can make three spades, that's entertainment. If you bet the structure will survive for a hundred years, that's engineering. See the difference?

—**Unknown Engineer**

A

Absorber A substance which attenuates stress energy.

Acceleration The rate of change of velocity per unit time (denoted as dv/dt or d^2x/dt^2) which can be specified in Grms or Gravity root mean square as is done for this book. This can be linear and/or angular motion. Its value is different for various bandwidths. 32.2 feet per second per second equals 1G.

Accelerometer An electromechanical transducer that converts acceleration into electrical energy. Two sizes are recommended for use in accelerated reliability testing, a standard size (threaded stud) and a miniature version for product monitoring. The units of acceleration are meters/sec/sec and this equates to Grms or Gravity root mean square, which is the area under the spectrum for a given frequency bandwidth.

Acceptable Quality Level (AQL) A measurement of the maximum percent defective for a sampling plan that can be considered acceptable. A low producer's risk should be considered in the plan and associated with the AQL.

Angular Frequency A/The circular motion of any of the three orthogonal vibration vectors, measured in radians per second.

Arrhenius Equation The equation describes the reaction rate of chemical reactions and migration effects. This equation is the basis for MIL 217 and when comparing the calculated reliability values to the actual field values, there is much disparity. Over the years, MIL 217 has undergone numerous revisions in attempts to reconcile these differences, all to no avail. Some have been successful in changing the failure values to reflect what their product experience in the field and these calculations have provided much more accurate MTBF estimates than the standard 217.

Auto Spectral Density (ASD) The average acceleration power in each band of an analysis. It is the same as Power Spectral Density (PSD). The square root of the area under the ASD curve is defined as the Grms or more commonly known as Gs.

B

Beta Testing Customer evaluation of a pre-released product so that final product deficiencies can be found and corrected prior to release. This is usually the first time in the product life cycle in which someone other than the producer has an opportunity to evaluate the product.

Broadband Signals (usually vibration) which are wide in frequency content and contribute to the overall acceleration measurement. Generally, systems with bandwidths of over 500 Hertz are considered to be broadband.

Burn-in A technique for accelerating the failures through the use of elevated constant temperature as described by the Arrhenius Equation. When component failure rates were high (up to about 1980), this process was successful but as component failure rates improved by orders of magnitude, this process became very ineffective at the product level or any other level. Typically, component manufacturers perform an extended burn-in at 150°C. Therefore, it is quite obvious why a 100°F product burn-in may be quite ineffective since most known failure modes are not activated below this latter temperature.

C

Charge Amplifier An amplifier that converts the charge signal from a transducer (usually an accelerometer) into a conditioned output signal.

Coherence The measure of similarity between two or more physical locations on a vibration table.

Corrective Action (CA) The implementation of a change intended to eliminate the source of the flaw in future production. Using a correct forming die might prevent a nicked lead. A bond flaw might be corrected by using a different pressure or perhaps by better cleaning. A defective solder joint might be corrected by using a different solder, improving paste deposition, a different temperature, or a different conveyor speed. Corrective action should be implemented only after the root cause of the flaw is clearly understood.

D

Dampening Material A substance that dissipates vibration energy. Same as absorber.

Degrees of Freedom The number of directions in which an object is free to move. An object movement in the vertical direction would only have one degree of freedom.

Detection Screen To detect in some manner that an abnormality exists, either electrically, visually, or by any other means. Note that an abnormality may be intermittent in nature and may only be observable under particular conditions, such as low temperature and/or low level or modulated six degrees of freedom vibration or some combination of stresses.

Device Defect Tracking (DDT) A process by which an abnormality is recorded in a database and driven through six distinctive stages of corrective action.[1] This is an extremely important process for any effective product reliability enhancement program to manage resource allocation. This is also known as Distributed Defect Tracking System (DDTS).

Design Ruggedization The process of finding the weak elements in a design and fixing them so that the design becomes very robust. It has the capability far beyond that which may be encountered in the expected field environment at no or minimal cost differential. Since the real field environment is rarely accurately known, large margins are therefore appropriate.

Destruct Limit (DL) A point at which the product ceases to function due to damage induced by a stress or a combination of stresses. Once the stress(es) is removed, the unit does not continue to function properly or at all.

E

Exponential Acceleration An acceleration in which the stresses are exponential or follow the power law.

F

Fatigue A condition in which the material has been weakened so that its strength can no longer withstand additional stress. This weakening can be a fracturing of the material or its atomic structure.

Failure Analysis A process by which accurate information regarding the failure mode of a component or assembly can be ascertained so that root cause is understood and corrective action can be implemented.

G

Grms The square root of the area under the Auto Spectral Density (ASD) or Power Spectral Density (PSD) curve or the square root of the mean square of the vibration signal. For a complete representation of the vibration signal, the Grms needs to be specified in conjunction with a bandwidth.

H

Highly Accelerated Life Test (HALT) A method used to uncover design and process related flaws that would otherwise go undetected until the product is in the customer's hands. It involves step stressing, rapid thermal transitions, as well as using combined stressing of the product under various environmental conditions.

Highly Accelerated Stress Audit (HASA) A statistical system designed in conjunction with a Highly Accelerated Stress Screen (HASS) that allows the user to audit the product's outgoing quality through a sampling of the product. The statistics will allow the user to conclude (with some risk) whether the product should be shipped if failures occur. Likewise, if no failures are detected (with some margin of error or risk), would the system have detected a failure had it occurred? The detection of a failure is analogous to the production process having gone out of control and not being detected by the production test process.

Highly Accelerated Stress Screen (HASS) The monitoring of the product's outgoing quality by applying the stress regimen to every product produced. When screening is not a requirement, an audit (HASA) should be the goal.

L

Latent Defect A defect that is not detectable because of too low a stress level or an incorrect stress which is, in effect, dormant.

M

Miner's Criteria The relationship that expresses the fatigue damage accumulation done in terms of a summation of a number of stress cycles at various stress levels.

Modulated Vibration The incremental stepping of six degrees of freedom vibration from a low level (possibly "tickle" or lower) to a limit which can be up to 50 percent of the destruct limit and then back down again, while dwelling at each step for a few seconds (15 to 30). This creates a "stair-case" going from minimum to maximum and then reversing itself.

Monitoring During stressing, this means that the product is powered on and is performing all or most of its intended design function while its output(s) are carefully being monitored for any abnormalities. Ideally, the monitoring is to be done by a computer or by links to intelligent modules.

Mean Time Between Failures (MTBF) The measurement of the mean or average time between a product's failures. If it is assumed that the failures follow the exponential model, then the reciprocal of the MTBF is the failure rate.

N

NAVMAT P-9492 Please see the seventeen page document, NAVMAT P-9492, for additional information. This was a proposal set forth by the U.S. Navy in May, 1979 as an aid to its suppliers for providing reliable products. It consisted of two major components, thermal assessment and vibration. The thermal information (see pages 3 through 8 in said document) was a survey of what some of the leaders in manufacturing were subjecting their products to. It provided misleading and ambiguous information since the reader had no idea what the nature and application of the product was that appeared in the survey. The vibration proposal (see 9 through 13 in the same document) was suggested with the technological limitations of the day. This vibration profile that many adopted had nothing to do with the end-use environment of the product and provided a false sense of security for those who subjected their products to it. It is not recommended that this document be used as a reliability resource.

O

Operating Limit (OL) A point reached during step stressing (or rapid thermal transitions) at which one or more of the product's specifications are no longer being met. At this point, one needs to determine if the operating margin is sufficient for the product. If the margin is not, then a redesign or correction of the fault needs to be made.

Operating Margin The margin that a product has beyond its published specifications which allow stresses to be applied and within which the product fully meets all of its specifications.

P

Patent Defect A defect that is active and detectable under correct conditions.

Precipitation The process of changing a flaw in the product from latent (undeveloped or dormant and usually undetectable) to patent (evident or detectable). An example would be to break (under stress) a nicked lead on a component or to fracture a defective bond or solder joint.

Proof of Screen (POS) An important phase of production screen development (HASS) provides the user with the confidence that the stress regimen does not remove an appreciable amount of the useful life from the product and that the stress regimen is robust enough to detect the flawed population in question. The normal HASS stress regimen is to be repeated for a minimum of ten times, preferably at least 30 to 50 times, without the product(s) failing or indicating any performance degradation.

Power Spectral Density (PSD) The average acceleration power in each band of an analysis. It is the same as Auto

Spectral Density (ASD). The square root of the area under the PSD curve is defined as the Grms or more commonly known as Gs.

Pseudo Load A thermal load that simulates the actual product's thermal load. For the sake of simplicity, the stress system control thermocouple that would normally be secured to the product during HASS or HASA is secured to this load simulator so that it does not need to be removed and secured prior to the execution of each profile.

R

Reliability Growth Test (RGT) A test in which the product's reliability is measured over time through the use of many units for a short period or fewer units for a much longer time. Usually, a reliability goal and failure criterion are also defined and set before beginning the test.

S

Sample Size A calculated sample size that is influenced by risks, the baseline failure rate and the shift to be detected in that failure rate. For this book, Acceptable Quality Level (AQL) type sampling will not be considered.

Screen Tuning The iterative process of refining and optimizing the stress regimen for higher throughput and demonstrated effectiveness.

Six Degrees of Freedom Vibration Vibration that provides non-coherent and non-stationary six axis vectors and thus, an almost random vibration profile. The six degrees of freedom consists of three orthogonal (or linear, x, y and z) and three rotational (or angular, pitch, yaw and roll) accelerations. Not all six degrees of freedom vibration systems are the same. Bandwidth, spectral shape and total energy, as well as energy in specific bands should be the key specification considerations when assessing these systems.

Shock Response Spectrum (SRS) An analysis method that predicts the peak acceleration responses of various frequency spring mass systems to an excitation.

Step Stressing A technique in which usually one stress is applied to a product being evaluated for a pre-determined time while the product's specifications are monitored. If the product is within its specification during the application of the stress, the stress is further increased in discrete, pre-assigned steps until the product's operating and destruct limits are uncovered. The product's specifications should *not* be derated as the stresses are increased unless they are specified in terms of that stress.

Stress for Life (STRIFE) STRIFE is the acronym for *STRess for LIFE*, a method used to uncover design-related flaws that would otherwise go undetected until the product is in the customer's hands. It may use sequential stresses. However, some use combined stressing of the product under various environmental conditions. This methodology is defined differently by each user, even within the same company and is not as effective as HALT because it lacks the combined effects of the various stresses and does not stress the product to the extremes that HALT does. This acronym was originally coined in 1971 at one of Hewlett-Packards' (HP's) Colorado facilities.[L]

T

Thermocouple A transducer that converts thermal energy to a low-level electrical energy. The thermocouple is created by mechanically joining two dissimilar metal wires (elements) together. A thermocouple can measure either air or product temperature. T-type thermocouples (constantin and copper) are the most commonly used in accelerated reliability testing. Polarity must be observed with the thermocouple wires and the connectors must be the appropriate type for the particular thermocouple being used. The failure to observe polarities will cause abnormal control and or monitoring system response.

Thermal Cycling Forcing the thermal chamber to repeatedly go from one temperature extreme to another, usually from a hot temperature to a cold temperature or vice-versa. Thermal cycling is beneficial in uncovering certain electronic defects and should be part of the overall stress regimen. High rate combined thermal and vibration systems can reduce the number of thermal cycles that are required to induce failures, compared to very low rate chambers.

Thermistor A negative temperature coefficient device used for limited temperature range measurements. These devices are not typically used in accelerated testing because of their limited high temperature limits. They also have a non-linear response that can be linearized with external precision resistors.

Tickle Vibration A low level six degree of freedom vibration that is used to uncover defects which would otherwise go undetected at high vibration levels. It is usually in the ten to two Grms range for most electronic products. It is empirically determined and varies from product to product.

Traditional Environmental and Qualification Methods Methods used throughout industry that, for the most part, were designed when electronic component failure rates were measured in percentages. Generally, these methods do not uncover the failures that are today measured in terms of Parts Per Million (PPM)—orders of magnitudes better. For design flaws that are seen across all or most of the population of units stressed, this method will work, but it will not precipitate many of the lower level flaws. These methods stress the product only to the level that meets the specifications or passes the test. They do little to improve the product reliability.

V

Verification HALT A HALT used to verify that the proposed corrective action has corrected the original defect and has not introduced a new failure mode. This HALT may be a complete or abbreviated HALT. When an abbreviated HALT is used (it's not really a HALT since it doesn't include all of the stresses) it uses the stress(es) under which the original defects were found. In other words, if the original unit failed under cold step stressing, the verification HALT may only include that stress.

References

1. Seusy, C. 1987. Achieving phenomenal reliability growth. *ASM International Proceedings* (March).
2. Henderson, G. 1999. Get a handle on fatigue to improve results from HALT-HASS stress screening machines. *Proceedings of IEEE 1999 Workshop*, 367–376.
3. McLean, H. 1992. Highly accelerated stressing of products with very low failure rates. *Proceedings of the Institute of Environmental Sciences (IEST)*, 443–450.
4. Hopf, A. M. 1993. Highly accelerated life test for design and process improvement. Proceedings of the *Institute of Environmental Sciences (IEST)*, 147–155.
5. McAfee, B. 1995. Application of accelerated testing in design and production. *Reliability and Maintainability Symposium (RAMS)*.
6. McKinney, K. 1995. Test cycle time reduction and improved quality by using accelerated testing techniques. *Proceedings of the Institute of Environmental Sciences (IEST)*, 72–79.
7. Oliveros, J. 1994. Application of using a pneumatic triaxial simultaneous vibration system in a production ESS environment. *Proceedings of the Institute of Environmental Sciences (IEST)*, 150–159.
8. Hobbs, G., and Mercado, R. 1984. Six degrees of freedom vibration stress screening. *The Journal of Environmental Sciences* (November/December).
9. Capitano, J. L., and Feinstein, J. H. 1986. ESS demonstrates its value in the field. *Proceedings of the Annual Reliability and Management Symposium (RAMS)*, 31–35.
10. Yoder, S. K. 1994. Shoving back—How HP used tactics of the Japanese to beat them at their game. *Wall Street Journal*, 8 (September).
11. Pugacz-Muraskzkiewicz, I. 1990. Arrhenius law in its application to the stress screening of electronic hardware—A model with a pitfall *Proceedings of the Institute of Environmental Sciences (IEST)*, 779–783.
12. Caruso, H. 1983. Significant subtleties in stress screening. Proceedings of the *Institute of Environmental Sciences (IEST)*, 154–158.
13. Wong, K. 1989. A new environmental stress screening theory for electronics. *Proceedings of the Institute of Environmental Sciences (IEST)*.
14. Smithson, S. A. 1990. Effectiveness and economics—Yardsticks for ESS decisions. *Proceedings of the Institute of Environmental Sciences (IEST)*, 737–742.
15. Hobbs, G., and McLean, H. 1994. Is uniformity and repeatability essential to vibration and temperature screening? *Sound and Vibration* (April): 22–23.
16. Studies done by the Technical Assistance Research Programs, Washington D.C.
17. Leonard, C. T. 1991. *Improved techniques for cost-effective electronics.* RAMS/IEEE.

18. Zunzanyika, K. and Yang, J. J. 1995. Simultaneous development and qualification in the fast-changing 3.5" hard-disk-drive technology. *The Annual Reliability and Maintainability Symposium (RAMS)*, 27–32.
19. Silverman, M. 1997. Summary of HALT and HASS results at an accelerated reliability test center. *Nepecon West Conference*, 866–892.
20. Cooper, M. and Stone, K. Manufacturing stress screening results for a switched mode power supply. *Proceedings of the Institute of Environmental Sciences (IEST)*, 133–139.
21. McLean, H. 1999. A method of estimating changes in product life from HALT—HALTPlus™. *Proceedings of IEEE 1999 Workshop*, 273–277.
22. Gusciora, R. H. 2000. Why continue to try proving that HALT really works for electronic parts? *Nepcon West Conference (March)*, 774–784.
23. Seusy, C. Reliability growth management in non-military industry. Hewlett-Packard Co.

Additional References

A. Hakim, E. 1991. Microelectronic reliability/temperature independence. *Quality and Reliability Engineering International* 7:215–230.
B. Wong, K. 1990. Demonstrating reliability and reliability growth with environmental stress screening data. Proceedings of the Annual Reliability and Maintainability Symposium (RAMS).
C. Wong, K. 1988. Off the bathtub onto the roller-coaster curve. Proceedings of the Annual Reliability and Maintainability Symposium (RAMS), 356–363.
D. McLinn, J. A. 1990. Constant failure rate—A paradigm in transition? *Quality and Reliability Engineering International* 6:237–241.
E. Wong, K. 1990. What is wrong with the existing reliability production methods? *Quality and Reliability Engineering International* 6:251–257
F. Pecht, M., Lall, P., and Whelan, S. J. 1990. Temperature dependence on microelectronic device failures. *Quality and Reliability Engineering International* 6:275–284.
G. Pastor, J. 1991. Selling new reliability concepts. *Proceedings of the Institute of Environmental Sciences (IEST)*, 293–296.
H. Leonard, C. T. 1989. How mechanical engineering issues affect avionics design. NEACON.
I. Leonard, C. T. 1991. Temperature, reliability, and electronic packaging. *National Electronics Packaging Conference.*
J. Leonard, C. T. 1991. Mechanical engineering issues and electronic equipment reliability: Incurred costs without compensating benefits. *Journal of Electronic Packaging* (March) 113:1–7.
K. Lambert, R. G. Fatigue analysis of multi-degree-of-freedom systems under random vibration, 43–71.
L. Bailey, R. A. and Gilbert, R. A. 1981. STRIFE testing for reliability improvements, *Proceedings of the Institute of Environmental Sciences (IEST).*

Trademarks and Service Marks

HASA is a process developed by the author while at Hewlett-Packard Company.

Delrin™ and ULTEM™ are trademarks of duPont.

HALTPlus™ and Virtual Sample Size™ are trademarks of AT&T Corporation.

IEST is The Institute of Environmental Sciences and Technology.

OVS™ is a trademark of QualMark Corporation.

QRS™ is a trademark of Screening System Inc.

HALT^{SM} and HASS^{SM} are service marks of QualMark Corporation.

About The Author

Harry McLean was employed by Hewlett-Packard (HP) Company for 25 years before leaving in 1993. During his career at HP, he held various positions in production, research and development (R&D), and quality and reliability engineering. His expertise includes extensive knowledge in the medical electronics field as well as personal printers. During his last five years at HP he was involved in reliability improvements for dot matrix impact and thermal inkjet printers. It was during this phase of his career that he uncovered a way through which the customer's field experience with defective products could be mimicked in-house by moving beyond the use of traditional environmental and qualification methods to HALT and HASS. While at HP, he designed and implemented the HASA process that is included in this book. Harry has written papers regarding this topic, as well as papers addressing computer programs to better manage production facility departments. He was with QualMark for four years following his time with HP, where he consulted with many companies in the successful application of HALT and HASS. While at QualMark, he was the project research and development manager. He also taught HALT and HASS seminars. Presently, he is Reliability Engineering Manager at AT&T Wireless Services in Redmond, Washington. Harry received his electrical engineering degree from Northeastern University in Boston, MA. He is fluent in Portuguese and has taught HALT, HASS and HASA techniques in Brazil. Harry has 2 patents regarding a fixture design and a mathematical model for deriving MTBF from the results of HALT. Both are expected to be issued shortly.

Index

Page numbers in italics indicate figures and tables.

A

Accelerated reliability programs, areas involved in, 120–121
Accelerated Reliability Test. *See* ART
Accelerated stress system block diagram, *88*
Accelerometer, 13, 21, 27, 93
Acoustical microscope, 98
Air ducting, 96
Alarm point, 61
Ambient conditions, 21–22
Application Specific Integrated Circuit (ASIC), 65
ARL. *See* Average run length curve
Array Technology, 23, 51
ART (Accelerated Reliability Test), 1–2
ASIC. *See* Application Specific Integrated Circuit
Auxiliary test equipment, 70
Average run length curve (ARL), 80–83

B

Bad news, 108
Baseline failure rates, 57–58, 72
Binomial distribution, 58–59
Binomial expansion, 70
Block diagram, *88*
Bottlenecks, 102
Burn-in, 2, 107–108, 126

C

Cables for HASS, 48–49
Caruso, Hank, "Significant Subtleties of Stress Screening," 39
Case study of present and proposed reliability programs, 101–112
  See also Selling new concept to management

Chambers, 70, 88, 95–97
  conventional, 123
Chatter, 65
Chlorine, 97
Cold and hot step stressing, 10–11
Combined vs. sequential stress regimens, 125
Comfort zone, 69
Comparing present and proposed reliability programs. *See* Case study of present and proposed reliability programs; Selling new concepts to management
Compressor systems compared to liquid nitrogen systems, 90–92
Consumer's risk, 72
Continuous Quality Improvement (CQI), 71
Control charts, 62–64
Control systems, 95
Coolant, 8
Cooling, 90
Corrective action, 2, 9–10, 11
  perspective on implementing, 15–20
CQI. *See* Continuous Quality Improvement
Critical value (CV), 59, 71–73
Cumulative defects, 74–78
CV. *See* Critical Value
Cycles to fail, 34–35

D

Databases, 2, 26
Defect detection, 44–45
Design of Experiments (DOE), 59
Design Verification Testing (DVT), 5
DeskJet printers HASA project, 55–67
Destruct Level (DL), 11, 14, 22, 122
Detection screens, 38, 40–41
Device defect tracking status description, *26*
Didreau, Deni, 1
DL. *See* Destruct Level

148   Index

DOE. *See* Design of Experiments
DVT. *See* Design Verification Testing
Dwells, 8, 10, 13, 14, 21, 22, 88, 90–92, 115, 117, 127

E

Einstein, Albert, 101, 113
Electrodynamic shaker, 23, 36, 94, 125
Employee training, 70
Environmental Stress Screen. *See* ESS
Environmental stresses, 4
Equipment for accelerated reliability testing, 87–99
 auxiliary equipment, operator safety, and ESD, 97–98
 chamber, 95–97
 control systems, 95
 failure analysis, 98
 temperature, 88–92
  comparison of liquid nitrogen systems and compressor systems, 90–92
  cooling, 90
  heating, 90
  OVS-2.5eHP (Qualmark Corp.), *89*
  QRS-600V (Screening Systems Inc.), *89*
  turbulence (uneven airflow), 89–90
 vibration, 92–95
Escape rates, 19
ESD safe environment, 97–98
ESS (Environmental Stress Screen), 1–2
 compared to HASS, 113–114

F

Failure analysis, 9–10, 65–66, 98–99
Fixture characterization, 42–44, 50
 See also Proof of screen
Fixture design, 50
FLT. *See* Fundamental Limit of the Technology
Franklin, Benjamin, 31
Frequency margining, 13
Fundamental Limit of the Technology (FLT), 14

G

Go-no-go indicators, 9
Good news, 108
Grms, 93–94

H

HALT (Highly Accelerated Life Test), 1–29, 32–33
 benefits and shortcomings, 3–4
 cold and hot step stressing, 10–11
 combined stresses in, 14–15
 compared to product qualification methods, 115–116
 comparisons of products with and without, 6–8
 and corrective action, 15–20
 and cultural change, 122–123
 defects by environment, *15*
 and designers' business concerns, 18
 example of temperature step stressing, *12*
 examples of success, 22–24
 frequency of use, 118
 graphical representation of DDT, *27*
 limits and issues of, 16
 margin discovery diagram, *9*
 number to perform, 119
 number of units to use, 121–122
 objective of, 117
 other HALT stresses and special situations, 13–14
 overview of, 2–6
 performance location, 121
 phases of, 3
 process of, 8–15
 product development with, *5*
 product levels performed at, 119–120
 product life with and without, *7*
 product limits with proposed statistical limits, *17*
 and products with short field life, 127
 purpose of, 116
 rapid thermal transitions, 11–12
 recording of failures and corrective action, 26–27
 and ruggedizing, 24–26
 savings using, *6*
 sigma and deviation impact on thermal limits, *19*
 step stress approach model, *10*
 summarized, 114–115
 summary of, 20–22
 thermal statistics diagram, *17*
 troubleshooting products under stress conditions, 27–28
 users survey, 123, *124*
 using the process, 17–19
 verification HALT, 15

Index 149

vibration, 19–20
vibration statistics diagram, *20*
vibration step stress, 12–13
HASA (Highly Accelerated Stress Audit), 2, 32, 55–87
  background of, 56–58
  control chart for, 62–64
  difference from HASS, 79
  examples of risks and Z values, *60*
  introduction to, 55–56
  monitoring system issues, 63–64
  objectives for a project, 57
  primary purpose of, 79
  problems uncovered through using, 64–66
  refined, 69–85
  sample size determinations, *61*
  statistical process overview, 58–59
  statistics system, 59–62
  two scenarios, 62
HASA (Highly Accelerated Stress Audit) refined, 69–84, *85*
  average run length curve, *82–83*
  background and assumptions, 70–71
  critical value (CV), 71–73
  cumulative defects, 74–78
  decision making flowchart, *85*
  equations, 71–72
  graphical tool for detecting defect level changes, 74–78
  HASA acceptance sampling plan, 81–84
  introduction, 69–70
  introduction to, 78–79
  operating characteristic curve, *83*
  probability of failures, *74*
  process flow, 79–80
  and product screening (HASS), 70
  sample sizes, *72*
  statistical application, 71–74
  typical lot acceptance sampling plan, 80–81
HASS (Highly Accelerated Stress Screen), 1–2, 4, 6–7, 23, 31–53, 69–70, 78–79, 82
  appropriate stress levels, 37–39
  cables for, 48–49
  compared to ESS, 113–114
  defect detection, 44–45
  detection screens, 40–42
  difference from HASA, 79
  and elimination of production steady state, elevated burn-in, 126

fixture characterization, 42–44
frequency of use, 118
ideal thermal profile, *41*
life determination in proof of screen, 45–47
precipitation screens, 39–40, 42
primary purpose of, 79
production product stress screen, 33–34
profile of, 44
proof of screen (POS), 42–47
rate of change of temperature, 36–37
screen tuning, 47–48
successes using, 51–52
summary of, 49–51
and vibration, 35–36
why it works, 34–35
Heating, 90
Hewlett-Packard (HP) DeskJet printers
  HASA project, 15, 23, 55–67, 70, 87
  See also HASA
Highly Accelerated Life Test. *See* HALT
Highly Accelerated Stress Audit. *See* HASA
Highly Accelerated Stress Screen. *See* HASS
Hot spots, 46
Hot step stressing. *See* Cold and hot step stressing
HP. *See* Hewlett-Packard

K

Kapton tape, 44
King, Jr., Martin Luther, 55

L

Laminar airflow, 89
Latent defects, 2, 39, 88
LDL. *See* Lower Destruct Limit
Liquid nitrogen, 8, 70, 88, 97
  compared to compressor systems, 90–92
Log-log paper, 36
LOL. *See* Lower Operating Limit
Lower Destruct Limit (LDL), 11, 40
Lower Operating Limit (LOL), 11

M

Magnavox, 23
Manpower, 70
Manufacturing Release (MR), 4

Materials engineering, 120
Mature products, 3–4
  definition of, 105
  and introduction, 108
Mechanically induced fatigue, 34–35
MIL-STD 105, 71
Motorola, 59
MR. *See* Manufacturing Release
MTBF calculations, 109
MTBF tests, 116

N
NDF. *See* No Defect Found
"New Environment Stress Screening Theory for Electronics, A," (Wong), 39
No Defect Found (NDF), 45
No Trouble Found (NTF), 41
Noise level (of chambers), 96
Non-relevant defects, 38–39
Northern Telecom, 51
NTF. *See* No Trouble Found

O
Oliveros, Javier, 52
Operating characteristic curve, 80–83
Operational costs of HALT and HASS for one year with two stress systems, *106*
Operator safety, 97–98
Orthogonal vectors, 12, 93, 115
Overhead reduction, 108

P
Passing the test, 122
Pastor, Joan, *Proceedings of the Institute of Environmental Sciences,* 101
Patent defects, 2, 39, 88
Planck, Max, 87
Pneumatic vibration system, 92–93
Polyvinyl Chloride (PVC), 49, 97
POS. *See* Proof of screen
Power cycling, 13, 65
Power Spectral Density (PSD), 93–94
Precipitation screens, 38–40
Probability of failures, *74*
Proceedings of the Institute of Environmental Sciences (Pastor), 101
Producer's risk, 72
Product accessibility (chambers), 95
Product development with HALT, *5*

Product life with and without HALT, *7*
Product limits with proposed statistical limits, *17*
Product qualification methods compared with HALT, 115–116
Production lot, 81
Production product stress screen, 33–34
  See also HASS
Proof of screen (POS), 42–47, 50–51, 93
Prototypes, 121–122
PSD. *See* Power Spectral Density
Pseudo load, 44
PVC. *See* Polyvinyl Chloride

Q
Qual testing, 102
Quotes, misguided, 122

R
RAM. *See* Random Access Memory
Random Access Memory (RAM), 65–66
Rapid thermal transitions, 11–12
Real-time system, 57
Reliability Enhancement Test. *See* RET
Reliability improvement programs promotion. *See* Selling new concepts to management
RET (Reliability Enhancement Test), 1–2
Revenue generation benefits, 107–108
RGT tests, 116
RGT/RVT, 107–108
Risks, *60,* 67, 72
Robust products, 32
Robustness, 2–3, 88, 94, 126–127
Root cause, 2, 9, 11, 15, 16, 104–105
Rotational vectors, 12, 93, 115
Ruggedizing, 14, 22, 24–26, 69

S
S–N (stress versus number of cycles to fail), *34*
Saddle effect, 93
Safety, 97–98
Sales support for chambers, 97
Sample size, 58, 71, *72, 74*
Sampling plan, 79–84
  HASA acceptance, 81–84
  typical lot acceptance, 80–81
Scanning electron microscope, 98
Screen design, 50

Screen tuning, 47–48
Seeding, 44–45
Selling new concepts to management, 101–112
  addressing potential management concerns, 106–108
  benefits of reliability improvement programs, 112
  comparison of on-going costs for present and proposed programs, *110*
  comparison of personnel and space requirements for present and proposed programs, *110*
  costs for proposed programs, *111*
  introduction, 101
  proposed program, 104–106
  the savings, 108–111
  See also Study of present and proposed reliability programs
  situation today, 102–104
  summary of savings between programs, *111*
Sequent, 51
Sequential vs. combined stress regimens, 125
Service by manufacturers of chambers, 96
Serviceability of chambers, 96
Shakers, 23, 36, 94, 125
  See also Six degrees of freedom
Sigma and deviation impact on HALT thermal limits, *19*
"Significant Subtleties of Stress Screening," (Caruso), 39
Simulation, 41
Six degrees of freedom, 8, 40, 92–95, 99, 115, 123, 125
  See also Vibration
Six-sigma, 59, 71
Smith, Hyrum W., 69
SMT. *See* Surface Mount Technology
SPC. *See* Statistical Process Control
Speed, 95
Spikes, 11
Spreadsheets, 2
Standing committee, 120
Statistical Process Control (SPC), 59
Statistical Tolerance Analysis, 59
Statistics for HASA process, 57–62
  application of, 71–74
  background and assumptions, 70–71
  characteristics for, 58
  control chart for, 62–64
  equation, 59–60, 66

risks, *60, 67*
sample size, 59, *61*
Step stress approach model, *10*
Step stressing, 10–11, 28
Stereo zoom microscope, 98
Stimulation, 41
Stress levels, 37–39
Stress for Life. *See* STRIFE
Stress screening. *See* HASS
Stress versus number of cycles to fail. *See* S–N
Stresses
  generic, 117
  product specific, 117
Stresses beyond design specifications, 116–117
STRIFE (Stress for Life), 1–2
Surface Mount Technology (SMT), 44, 93–94
Surface mount transistor lifting and tearing up substrate, *37*
Synergism, 38, 116
System capabilities of chambers, 96–97

T

Tables, 93–95
  See also Vibration
Taguchi, 71
Teflon, 49, 97
Temperature deviation, 16
Temperature ramp rates, 125
Temperature step stressing, *12*
Tensile stress, 34–35
Test engineering, 120
Test to spec, 105
Thermal cycling, 21, 36–37, 92
Thermal limits, *19*
Thermal profiles, *41–42*
Thermal specifications, *16*
Thermal statistics diagram, *17*
Thermal transitions, 11–12
Thermocouples, 10, 20, 44, 95
Throughput, 116
Tickle vibration, 13, 22, 40
Total Quality Control (TQC), 71
Total Quality Improvement (TQI), 71
TQC. *See* Total Quality Control
TQI. *See* Total Quality Improvement
Traditional product development, *5*
Traditional thermal profile, *42*

Turbulence, 89
Turbulence (uneven airflow), 89

U

Ultimate Technology Corporation, 24
UOL. *See* Upper Operating Limit
Upper Operating Limit (UOL), 11, 13, 40
User configurable, 95
User friendliness, 95
Utilities, 70

V

Verification HALT, *15*
Vibration, 92–95
  conventional table, 123
  and HASS, 35–36
  tickle, 13, 22, 40

Vibration deviation, 16
Vibration statistics diagram, *20*
Vibration step stress, 12–13
Voltage margining, 13

W

Warranty reports, 31–32
Wearout mode, 6–8
Wong, Kam, "A New Environment Stress Screening Theory for Electronics," 39

Z

Z values, *60, 66*